Peter Schwartz is the co-founder and Chairman of Global Business Network, one of the world's most eminent research and consulting firms. He advises a variety of blue-chip clients on scenario planning, from government agencies to giant corporations including the CIA, Boeing and Texaco. He is the author of *The Art of the Long View*, a bestseller, and co-author of *When Good Companies Do Bad Things* and *The Long Boom*.

ALSO BY PETER SCHWARTZ

The Art of the Long View:
Planning for the Future in an Uncertain World

When Good Companies Do Bad Things:
Responsibility and Risk in an Age of Globalization
(with Blair Gibb)

The Long Boom:
A Vision for the Coming Age of Prosperity
(with Peter Leyden and Joel Hyatt)

China's Futures
(with Jay Ogilvy)

PETER SCHWARTZ

INEVITABLE SURPRISES

THINKING AHEAD IN A TIME OF TURBULENCE

First published in Great Britain by The Free Press in 2003
An imprint of Simon & Schuster UK Ltd
A Viacom company
This edition published, 2004

1 3 5 7 9 10 8 6 4 2

Simon & Schuster UK Ltd
Africa House
64–78 Kingsway
London WC2B 6AH

www.simonsays.co.uk

Simon & Schuster Australia
Sydney

A CIP catalogue record for this book is available
from the British Library.

ISBN 0-7432-3911-3

Printed and bound in Great Britain by
Bookmarque Ltd, Croydon, Surrey

FOR CATHLEEN,

who never fails to surprise me

CONTENTS

PREFACE

Our world has become no less surprising in the year since the hardcover edition of this book went to press, and our capacity for the denial of those surprises has become even more obvious.

In March of 2002 the United States invaded Iraq with massive force only to be surprised by the swiftness of the Iraqi military collapse. But that was not the last surprise awaiting the American forces. First came looting so pervasive and intense that the ability to rebuild Iraq vanished along with the miles of copper wire and plumbing stripped from buildings. The looters took computers and all the hardware necessary to operate a modern society, along with the historical treasures of an ancient nation. Then came the daily killings of American soldiers and relief workers by the remnants of Saddam Hussein's forces in a sustained and low-level guerrilla campaign. As I write this, the 500th American soldier has lost his life. A brilliantly-executed air and ground campaign destroyed the horrific regime of Saddam Hussein but unleashed devastating surprises for which the Americans were apparently unprepared. The war may have been won (though that is not a sure thing as the insurgency continues), but so far there has been little victory in the peace.

In the chapters that follow, the reader will find that we are often surprised by major events that in hindsight were obviously inevitable, and equally important, foreseeable. The dominant intellectual strategy that people bring to bear on the future is denial. This book aims, however, to penetrate the denial by exploring how to anticipate some of what lies ahead, and what some of the most important surprises are likely to be. The final chapter envisions a world of maximum surprise, and how we can respond today to the possibility of such a future.

The past decade has seen remarkable events and changes:

The economic takeoff of China and India, and the depression in Japan

The birth of the network economy and the boom/bust that resulted

9-11 and the War on Terrorism

The war in the Balkans

The discovery of dark energy

The always-on, always-connected, always-available society

The *Columbia* disaster and a Chinese taikonaut in orbit

The creation of the Euro

The rise of the antiglobalization movement

The collapse of corporate giants like Enron and Parmalat

The acceleration of productivity

The Y2K non-problem

The movement of nanotechnology from the fringe to the center of science and technology

The death of the music industry as we knew it

And many more . . .

Are these just random events concatenated one after the other, with no meaning or order? Or is there a discernable pattern behind them that just might let us anticipate some of the inevitable surprises ahead? Obviously I believe that these kinds of events are the outcomes of deeper forces at work.

Since the end of the Cold War, two great forces have been expanding the circle of prosperity and common interests. The first force has been the ability of more nations and more people to capture the benefits of knowledge-driven growth made possible by new technology, investment, and education. The second major force has been better governance. By this I mean more competent, less corrupt, more open, and sometimes even more democratic governments. The two best examples of this are, of course, China and India, following in the footsteps of Japan, Korea, Taiwan, and the Western industrialized nations before them. The global and explosive nature of this new growth is of course made possible by the networked information-technology revolution. Knowledge spreads nearly instantaneously around the world and high rates of productivity growth are made increasingly possible.

But there are at least four countervailing forces, and it is at the intersection of these forces that the surprises are likely to occur, for good or ill. First and most difficult is the challenge of radical Islam. Will the radical and violent fringe of Islam, whose vision is carrying their fellow Muslims back to the caliphate of the 13th century, dominate the Islamic world? Or will the Islamic countries join other nations riding the waves of knowledge-driven growth and better governance? Today many of their nations are failed theocracies. And even those that are wealthy because they are resource-rich, such as Saudi Arabia and Indonesia, are riven by the struggle between a retrograde vision of the future and modernization.

The very successes of economic growth are beginning to push up against the limits of the Earth's carrying capacity. It is this tension between economic growth and the environment that is the second major force that stands in the way of a more prosperous future. Climate change and the potential for a rapid transformation of the Earth have become even greater concerns recently. I happened to lead two major research projects on the issue in the last year. One conducted for the Pew Center on Climate Change focused on the question of what it would take for the United States to lower the trajectory of growth in greenhouse gases, especially CO_2. The study found that even under the most optimistic technological assumptions, including the continued development of new technology such as hydrogen fuel cells and carbon sequestration, unless there is government intervention, such as a limit on carbon emissions with trading permits, we will be much worse off in two decades. The second study was conducted for the Office of the Secretary of Defense and focused on the possible impacts of a period of abrupt climate change: rapid warming leading to even more rapid cooling, especially in the northern latitudes. The conclusions of this study suggested that not only were there early signals that such a change was coming, but that the consequences could be enormous. The result could be a much colder, drier, windier world struggling over water, food, and energy.

The third major force is also the result of success or at least partial success. More and more people are enjoying the benefits of prosperity. Over the last decade in India and China, at least two- to three-hundred-million people have moved out of poverty. However, at the same time, billions in Africa, Latin America, the Middle East, and Central Asia have fallen further behind. Globalization has worked near miracles for many, but even more are fearful of being left behind. The widening gap between those out on the frontier of the wave of knowledge-driven growth—the new Indian software entrepreneurs and the new Chinese industrialists—and those whose skills leave them ill prepared to play the new game has led to tension, protest, and a mounting push back against the integrated global economy. Farmers in the wealthy world persist in policies to protect their prices at the expense of the farmers of the developing world. Round after round of global trade talks, from Seattle to Cancun, have collapsed because of these conflicts between the haves and the have nots. The potential for protectionism looms large.

The recent SARS epidemic and the possible outbreak of Avian flu in Vietnam are a hint of what is to come. The economic disruption of global and regional plagues is the fourth major force pushing back against the future. Singapore experienced the worst economic crisis in its history as a result of SARS. Output fell at a seven percent annual rate as everyone stopped coming to Singapore. Empty Singapore Airlines airplanes and vacant hotels were all the result of SARS, occurring on top of the impact of terrorism, especially the Bali bombing. These new diseases are not the last. A number of conditions are leading to new and virulent diseases emerging from dense human societies interacting with the environment in new ways, and these diseases can spread nearly instantaneously around the world via the airline system. Until we find a way to counteract them they will continue to arise and continue to disrupt the global and regional economy.

So the context in which the inevitable surprises ahead are developing is one of a world driven forward by knowledge-driven growth and better governance, and pushed back by the violence of radical Islam, the strains against the Earth's carrying capacity, the social stresses of wealth vs. poverty, and the spread of deadly diseases. With that context in mind, it is possible to look ahead to some of the surprises that lie just over the horizon.

If the positive forces are able to manage the resistance created by the disruptive forces, then we can see some beneficial surprises ahead:

1. China will continue to grow internally as its markets expand and it becomes the world's industrial heartland, and it may also take its neighbors along with it as they all become realigned around China in what might come to be called (a little tongue in cheek) the Greater China Co-Prosperity Sphere.

2. An even bigger surprise will be the accelerating take-off of India. Service industries and software linked to the world via high bandwidth communications are leading to new prosperity, at least in some parts of India. The knowledge of the English language, the rapid growth of the educated population, and the connection to the success of Indians in the U.S. high-tech industry are the competitive advantages that have enabled the new growth dynamic to take off. The recent peace overtures with Pakistan are almost certainly linked to this new prosperity. Indians now have more to lose, and war, or even the looming threat of it, could seriously undermine these new hopes.

3. When the boom of the nineties went bust in the spring of 2000 many people wondered whether the growth was merely a one-time occurence due to a fevered Internet mania, or something more fundamental. The intervening years have answered that question. The big surprise has been how robust productivity growth has been. It has been so strong that even as economic growth has come back following the recession, job creation has been very modest. There is now high output with little need for new labor due to high productivity driven by networked information technology. The technology continues to advance and productivity continues to grow and, as has usually happened in the past, as demand grows, eventually the job-creation engine kicks in. So the surprise ahead is that even as foreign white-collar competition grows and productivity remains strong, new jobs will be created in large numbers across many industries and services, both new and old.

4. A huge geopolitical surprise could be a "new" George W. Bush in his second term, if he is reelected. (As this goes to press that remains uncertain.) Pegged as a unilateralist president in his first term, with little willingness or ability to cooperate with other nations, he could change his direction in his second term. If the violence in the war with radical Islam is sufficiently contained, he may come to see the virtues of building a stable global system capable of integrating the Islamic nations into the modern world. Having laid waste to the old institutions, a new set focusing on security, development, the abatement of global disease, and the protection of the environment might result.

It is quite plausible that the disruptive forces could overwhelm the spread of knowledge and even the best governments. If so, a different set of surprises could lie ahead:

1. The war on terror could spread, engulfing nations such as Pakistan, Indonesia, and even Nigeria. The horrors of weapons of mass destruction could even come into play as chemical, biological, or even nuclear weapons are employed in the name of Islam. No weapons of mass destruction were yet found in Iraq as this goes to press. But there is strong evidence that they, especially chemical weapons, were there. Were they smuggled out before the war and are they now possessed by radicals somewhere else in the world?

2. We are likely to see significant new regional conflicts that could even be so large as to have global consequences. Wars in Africa are inevitable but not likely to be globally significant. Turbulence in the Caspian region appears increasingly likely, and with its increasing importance as an oil- and gas-exporting region, the impact of these local conflicts could be very great indeed. But the most worrying is the mounting internal tension in Saudi Arabia. If today's highly-problematic royal regime were replaced via a revolution in the streets led by the radical Islamics, as happened in Iran in 1979, then the outcome would be a similar radical Muslim regime hostile to the West and to America in particular. Of course the greatest issue in this scenario is the importance of Saudi Arabia's oil reserves. Would the United States let the world's largest pool of oil end up in hostile hands? Probably not, and the result would be Gulf War III.

3. We have already begun to see higher energy prices, especially for natural gas and oil. Gas prices have tripled in recent months. Demand continues to rise but supply has not kept up. The limits of Canadian natural gas have been nearly reached, and the United States has prevented the building of liquefied natural gas facilities, which has stopped imports from more distant sources. Exploding demand for oil in China and political problems in the Gulf, Nigeria, the Caspian region, and Venezuela have led to demand outpacing supply, resulting in persistently high prices, beyond even the desires of OPEC.

4. An outbreak of protectionism between the United States, Europe, and China, or between rich farmers and poorer ones, could be worse than sand in the wheels of global growth. It could put them in reverse gear and even drive the world into a depression as catastrophic as the worldwide economic decline of the 1930s.

5. And finally, the evidence of abrupt climate change could become so convincing that the world would have to gear itself up rapidly for a future that would put enormous strains on our ability to supply seven billion people with food, water, and energy. The costs and tensions of that effort would fundamentally reshape the economic and political trajectory of the future.

Many of these surprises, and others, will be explored in the coming chapters. History provides ample reason to believe that we can expect inevitable surprises ahead.

CHAPTER 1

Inevitable Surprises

In a world of surprises, what can we count on?

As I write this, in early 2003, the question has never seemed so relevant. Some have lost their life savings in the economic turmoil of the last few years. Others have seen promising jobs and businesses abruptly disappear. A few of us have lost family members or friends in terrorist attacks. Many of us took for granted the idea that the United States, as the most powerful nation in the world, was virtually unassailable by outsiders. Nineteen madmen shattered that illusion on a September morning in 2001. Has anyone seriously believed, since then, that New York, Washington, or other American cities will never be in danger again?

Leaders of organizations, from corporations to government agencies to nonprofits to unions, have seen our financial and market assumptions overturned. High-growth new media enterprises that were about to reshape the worlds of communication and retail suddenly went bankrupt. Time Warner and AT & T may well suffer similar fates to Enron and WorldCom. Latin America's economy abruptly entered freefall last year. Fourteen million African children have been orphaned by AIDS, a disease that nobody knew existed thirty years ago. The global middle class is burgeoning in, of all places, the immense socialist societies of China and India.

You may be reading this book a year hence, or five years, or perhaps more, when these specific stories will seem like ancient history. Yet the fundamental point remains: You, too, live in a world of surprises. For surprises are the norm. There will be many more moments to come when the assumptions you've lived by sudden fall away—inflicting that same queasy feeling you get when an elevator drops a little too suddenly, when an airplane hits an air pocket, or when a roller coaster moves past the top of the curve and lurches into its descent. There will also be beneficial surprises to come—when impossible, unthinkable opportunities and technologies suddenly become real, for you (or someone else) to cultivate, develop, and use.

Historically, upheaval is not a new condition. To be sure, there have been some relatively surprise-free centuries in human history; life for most people in medieval Europe was much the same as it had been for their parents. But since the scientific discoveries of the seventeenth century, complexity and turbulence in the world at large have been facts of life, looming larger and larger in people's concerns until today there is hardly anyone unaffected by them.

At the same time most of us still feel emotionally that things *should* be stable and certain; that once we're over the next hump of crisis, life will naturally return to tranquil normalcy. And there are things we don't want to see strapped into a roller coaster: Our country's security. Our companies and jobs. Our retirement accounts.

Is there a better way to live with this tension than just to hang on for the roller-coaster ride and react to every new surprise thrust at you? Yes, there is. There are still certainties—still facts and factors that we can rely on and even take for granted. For example, the quality of the natural environment—of air, water, and land use—in the industrialized world will significantly improve over the next thirty years. The use of "soft power" (moral suasion) will be more and more influential in diplomatic and military arenas, even as "hard power" (weapons and military technology) grows more prominent in the American federal budget. And the economy will revive: not in the same bubbling, effervescent form as it took in the late 1990s, but in a form that makes general prosperity seem once again accessible.

There are many things we can rely on, but three of them are most critical to keep in mind in any turbulent environment.

> **First:** There will be more surprises.
>
> **Second:** We will be able to deal with them.
>
> **Third:** We can anticipate many of them. In fact, we can make some pretty good assumptions about how most of them will play out.

We can't know the consequences in advance, or how they will affect us, but we know many of the surprises to come. Even the most devastating surprises—like terrorist attacks and economic collapses—are often predictable because they have their roots in the driving forces at work today. On September 11, 2001, we saw the tragic consequences of ignoring those predictions. The terrorist attack that day was perhaps the most forecast event in history. A half dozen reputable commissions over the last twenty years had suggested that an incident very much like this might occur. Many predictions had singled out the World Trade Center (in part because it had been attacked once before), mentioned the use of airplanes as weapons, or specifically referred to Osama bin Laden. No one knew when the event would happen—it might be next week, it might be two years from now—but the details were foreseen. Yet most people, in both the Bill Clinton and George W. Bush administrations, focused their attention elsewhere prior to September 11: on domestic priorities, campaign priorities, and other military arenas including missile defense.

A few people in responsible positions did look ahead. For example, following the surprising end of the Cold War, the President and the Congress created a commission chaired by Gary Hart and Warren Rudman, to advise it on a new fundamental national security strategy. I led the scenario team for the Hart-Rudman Commission. Our report, released a few months after George W. Bush was inaugurated in 2000, warned that terrorist incidents represented the greatest threat to the United States. In one scenario we anticipated terrorists destroying the World Trade Center by crashing airliners into it. Our most urgent recommendation was that the U.S. needed new levels of capability in homeland defense.

The Commission's work, and other similar efforts by various critical agencies, did not prevent the attacks, but they did contribute to the decisive speed and competence with which the U.S. responded, especially in the first few months.

In the coming decades we face many more inevitable surprises: major discontinuities in the economic, political, and social spheres of our world, each one changing the "rules of the game" as it is played today. If anything, there will be more, not fewer, surprises in the future, and they will all be interconnected. Together, they will lead us into a world, ten to fifteen years hence, that is fundamentally different from the one we know today. Understanding these inevitable surprises in our future is critical for the decisions we have to make today—whether we are captains of industry, leaders of nations, or simply individuals who care about the future of our families and communities. We may not be able to prevent catastrophe (although sometimes we can), but we can certainly increase our ability to respond, and our ability to see opportunities that we would otherwise miss.

The global financial networks provide an ongoing example of this. Errors and panics continue to occur, but the financial world learns from these disasters. The 1929 financial crisis led to ten years of global depression; the 1987 financial crisis, which was arguably a greater calamity judged just from the lost market capitalization, led into a completely different outcome—a mild recession and then a boom. One reason for the difference: the financial institutions and regulators learned something from 1929. They are still learning. They will make many mistakes in the future, but they will not permit precisely the same kinds of abuses of research and accounting that led to the stock market crisis of 2000–2001.

The Nature of Predetermined Elements

How do I know all this? Because I have one of the most interesting jobs in the world. I lead Global Business Network (GBN), the world's preeminent research and consulting firm focused on scenario planning. I am also a venture capitalist—a partner at Alta Partners, one of the oldest and largest venture-capital funds. I am also occasionally invited by filmmakers to help them develop the details and plot lines of their films about the future—most recently with Steven Spielberg in the feature film *Minority Report*, which was released in 2001 and set in the year 2050.

With my GBN hat on I advise major corporations and leading governments on long-term decisions. I help them look ahead and figure out today's actions based on long-term perceptions and insights. I help them see the big surprises, and the driving forces that are shaping the potential futures that may lie ahead. I help them see what is inevitable and where the fundamental uncertainties lie. Then, as a venture capitalist, I place large bets on the future, helping to create the potential tomorrow that I would like to see emerge. Finally, in my motion picture advisory role, I help imagine the consequences of these large-scale trends for the day-to-day lives of ordinary people.

In all three roles I have become increasingly aware of the critical forces that will affect the world, in ways that most decision-makers do not automatically expect. These forces are what scenario planners call "predetermined elements": forces that we can anticipate with certainty, because we already see their early stages in the world today. We know they are inevitable because they have already begun to take place. They are also going to surprise us because, while the basic events are virtually predetermined, the timing, results, and consequences are not. We do not know exactly how these events will play out, or precisely when they will occur. But we can anticipate the range of possible results, and the ways in which the rules of the game may change thereafter.

Scenario-planning exercises, such as the kind I conduct at Global Business Network, often include detailed examination of these kinds of "predetermined elements." Indeed, one of the great innovators of scenario practice, Pierre Wack, used to make them the linchpins of his scenarios for Royal Dutch/Shell in the 1970s and early 1980s. He knew, after intensive study and thought, that inevitable surprises were coming down the pike, and that Shell's success in a turbulent marketplace depended on the company's ability to pay attention to them in advance.

Pierre used to compare his work to the prediction of floods on the Ganges River in India. "From source to mouth," he would say, "the Ganges is an extraordinary river, some fifteen hundred miles long. If you notice extraordinarily heavy monsoon rains at the upper part of the basin, you can anticipate *with certainty* that within two days something extraordinary is going to happen at Rishikesh, at the foothills of the Himalayas." Three days later, he would add, one could expect a flood at Allahabad, which is southeast of Delhi; five days after that, one could expect a flood in Benares, at the river's delta. "Now, the people down here in Benares don't know that this flood is on its way," he would conclude, "but I do. Because I've been at the spring where it comes from. I've seen it! This is not fortune-telling. This is not crystal-ball gazing. This is merely describing future implications of something that has already happened."[1]

The same is true for the inevitable surprises in this book. Chapter 3, for example, describes the waves of population migration that are bound to change societies throughout the world in the next two decades. In the United States, English-speaking descendants of Western Europeans will find their majority yet more reduced—which means Americans laws, institutions, and culture is about to undergo a sea change. European leaders will struggle with their own influxes of refugees and Islamic immigrants for years, if not for decades; and China may well face similar issues in Asia. To write this is not fortune-telling; the factors which make these waves predetermined have been visible for years. We've seen them! To be sure, the outcome of those migrations is not certain; but the success of our businesses, our governments, and perhaps even our life choices depends on being able to distinguish the aspects that *are* certain, and to act accordingly—even if it feels uncomfortable to do so.

This idea for this book was conceived in mid-2001, when I was contacted by Robert Rubin, the former secretary of the treasury under Bill Clinton, and now vice chairman of Citicorp. "We keep getting surprised by big things," he said, "whether it's debt crises in Brazil and Southeast Asia, or busts in the stock market. I'm gathering my advisory board, and my senior management together for a couple of days. Tell us what the big surprises are going to be. We want to avoid them."

At first I was wary. The problem with predicting the future, as every prognosticator knows, is that one's mistakes are remarkably visible in hindsight. But as I researched the possible trends and forces that might affect the future of Citicorp, I realized that many of them were not only plausibly likely to take place, but were predetermined to do so. When I went to deliver the talk, I was amazed to discover that they already had a sense of most of what I had to tell them! Each of them knew some of the facts I had to offer. None of them had put it all together—either separately or in groups—to make sense of the whole story. And thus they kept on getting surprised. But as I spoke, each kept nodding yes, as if to say, "Of course."

There was a lot that Citicorp could learn about the surprises of the future. The facts were not the issue; like most people in responsible positions in significant businesses, they were aware of the facts. But they could not always put them together to see their consequences. And the same is true for the rest of us.

Denial and Defensiveness

If the future is so predictable, why do so many businesses and organizations have difficulty putting the facts together? One would think that many people would be well practiced by now, for discontinuities have been a regular fact of life since at least the mid-1960s. (Consider: the Kennedy assassinations, the oil shortages of the 1970s, "stagflation," the end of the Cold War, the shifts in medical and communications technology, and the impact of climate change, just for starters.) Intellectually, it's easy to recognize that some of our working assumptions have been wrong, that we're on a roller coaster of events that puts our organizations and our livelihoods at risk, and that we have to be prepared. But doing something about it is another matter entirely.

When an inevitable surprise confronts us, there are two different types of natural reactions. Both of them can lead to poor decision making.

The first is denial—the refusal to believe that the inevitabilities exist. This was one of the key reasons, of course, why the U.S. government was unprepared for the attacks on September 11, 2001. Enough people in positions of authority simply refused to believe the need was great and urgent enough to justify rethinking the structure of our national security system. When in denial about an inevitability, people tend to blithely act as if it didn't exist, and as if there were no need to break from routine and prepare for it. The losses that result can be immense.

Similarly, one can have a great deal of sympathy for the employees of Enron who saw their stock portfolios and pension funds collapse in late 2001. It was beyond their control to try to salvage their pensions, and even the cautious among them were probably devastated. But one might also recall that on the way up, when the stock price was rising because of false profit reports, these people were enjoying it, denying that there was a potential for a crash, and in many cases reminding others less fortunate around them of their own good fortunes. A little foresight and reflection ought to have suggested that a price that goes up so quickly can come down just as quickly, and that—whether or not Enron's management supported it—perhaps they ought to diversify their holdings. To the extent they thought, *Nothing bad can happen to this company because we're growing so fast,* they were deceiving themselves. (Indeed, many Enron employees *did* take note of danger signals and diversified in time, thus weathering what would become a very difficult storm.)

Denial is perhaps the most dangerous response one can make when evidence of an inevitable surprise presents itself. Today, many political leaders are in denial about several of the surprises described in this book: global climate change, the inevitability of new diseases, and the dangerous "hot spots" of Mexico, the Caspian Sea, and Saudi Arabia. In Europe, denial of the realities of migration may tear the continent apart.

The second natural reaction to any turbulent crisis is defensiveness. This is a kind of opposite to denial. People take the inevitable surprise so seriously that they freeze; in their minds there is no viable way to act except to find a safe place, hunker down, and wait for it to all blow over. They reduce their investments and activities, focus on their immediate and narrow self-interest, and wait for another stretch of relative calm to set in before they are ready to take risks again. If they are corporate leaders, they cut costs and innovation. If they are political leaders, they look for short-term gains.

On a visceral level this defensive response makes sense. Emotionally, you feel much less in control of your destiny than you thought you were. (Your actual level of control is the same as it always has been, but it doesn't *feel* the same.) Maybe you can't control some external forces, you reason, but you can at least minimize your exposure to them.

Unfortunately, this strategy also tends to produce poor results. You are making one of the riskiest moves of all: to do nothing in the face of uncertainty. For example: If you lost money in the bursting of 2000's stock market bubble, it's natural to think: *I won't invest in corporate stocks ever again, because they can't ever be trusted.* But not all company leaders are as shortsighted (or exploitive) as those of WorldCom and Enron were. The bursting of a bubble is not a signal to stop investing in all companies—or even in technology companies, trading companies, or "new-economy" companies. It *does* mean that your criteria for investment need to be tougher. Your homework and due diligence need to be more sound. Rather than avoid taking risks, you need to be smarter about your risks.

While being emotionally understandable, both denial and defensiveness are fundamentally irresponsible, especially on the part of corporate leaders. Unintentionally, they foster an attitude of victimization throughout the organization: "There was nothing we could have done about our poor performance. We were overwhelmed by events." The airline industry has been a prime example of this: "There was a terrorist attack and demand for long-distance travel dropped, so we couldn't make our profit targets." Well, if you're in the airline business and you don't have a contingency plan for expanded terrorist activity, then you're probably going to suffer sooner or later anyway.

This book is for people who want to get past denial and defensiveness, to be the masters of their own fate in a world full of surprises. The first step in making that transition is to pay attention to the inevitable surprises of the future, and to develop strategies for dealing with them.

Different aspects of the environment demand different strategies. We know about some of the surprises in this book, for example, precisely because they have been a long time coming. They move in slow, steady ways; they can be seen coming for decades. The population slowdown falls into this category; so does the continued evolution of the computer and the impending global climate change. There is a long time to prepare for them—which is good, because they will require a long time of preparation.

Other surprises are both enormous in their implications and enormously abrupt. Everything is different after they take place. The release of Nelson Mandela from prison in South Africa was like that; many people inside the country were poised for apartheid controls to grow more strict. The end of the Cold War and the collapse of the Soviet Union was another. The fall of the Japanese banking system was a third; in the 1980s we were all prepared for "Japan, Inc." to virtually dominate the United States. Other examples are the takeoff of the Internet, the Asian financial crisis, the stock market boom and bust, and, of course, the attacks on September 11, 2001.

Such system-changing, disruptive events are far more common than most people imagine. We are getting at least one per year on a global level these days. And yet most businesspeople behave as if they live in a continuous environment, as if their business plans and projections are going to be relatively linear. In a world in which there are regularly occurring upheavals, which will fundamentally change many basic assumptions about the way the world works, the most effective strategy is conscious resilience: balancing short-term reactions with long-term vision, and putting in place the necessary preparations so that you can rapidly shift direction if need be.

IBM in the 1990s provided a good example of exactly this kind of navigation. For decades the company had a basic business model: leasing mainframes and providing ongoing support. They developed the IBM PC, the best-selling personal computer of the early 1980s, but never saw that as central to their business. It was peripheral to the mainframe structure that they had developed for their customers.

Then everything changed. Apple introduced more user-friendly computers that ate into their market share. They lost the battle with Microsoft for control over the personal-computer operating system. Their customers lost interest in the IBM mainframe and client-server structure as computers became cheap and powerful enough to operate in a more networked, interdependent fashion. And they began to watch the rest of the computer business move out from proprietary systems onto the World Wide Web and the Internet. All they had left was a large amount of cash, and a solid reputation—admittedly two enormous assets, but only if they deployed them.

They moved rapidly in the early 1990s to change their business model. IBM suddenly became the world leader in providing outsourced computer services, a field that EDS and Perot Systems had dominated in the 1980s. To accomplish this IBM had to bypass their world-renowned computer sales force, and essentially invent a consulting business from the ground up. The consulting business then drove their services business, which went from zero to $30 billion in revenues in less than a decade. This allowed them to shift their manufacturing orientation—to focus on the hardware needed for the services they provided. Few companies would have been able to make that move so quickly.

One reason that IBM *could* move so quickly is that they were attuned to all aspects of the computer business. Though the personal computer had been a sideline for the company, their experience with it gave them a gut feel about the changing marketplace and business environment that no other computer mainframe company had.

Xerox, for instance, which was also a market leader in the 1970s, was blindsided and almost bankrupted by its failure to act in the 1990s. They knew about the changes that were occurring in the marketplace; in fact, in 1995, GBN had conducted a scenario exercise with their senior management, in which one of the scenarios was called "The Death of Xerox." When we presented it to them, they said, "We would never do any of those things you have us doing." Then, over the next few years, they systematically made every move we had warned them against making in that scenario. Today Xerox is a much smaller printer and copier company.

There is a twofold process embedded in the information in this book. First, it seeks to understand the kinds of inevitable surprises that lie ahead of us, particularly in the next twenty-five years. That is the period of time in which most of the enterprises starting up today will come to fruition, and in which most of the established corporations (and governments) will have to reinvent themselves. Second, it suggests the kinds of steps that would allow a company or organization to thrive, given the inevitable surprises awaiting us.

Sometimes you may be able to influence the outcome of a surprise. If it's a good thing, you can make more of it happen, and if it's a bad thing, you may be able to prevent it. Sometimes you may be able to take advantage of it, because you have developed relationships, products, financial resources, and information that put you in the right position when the time comes. Sometimes, if you see a big surprise coming and are convinced it will happen, you can act with confidence in the face of apparent risk that others judge to be high—but that you know is lower than it seems, because you've thought about and understand the surprises. And you can also make sure you have the resources to weather the storms that you see coming—particularly the financial strength that allows you to get through crises without shuttering parts of your business.

One common question is: Can you actually see these things coming? I'll let you judge for yourself, by exploring the "face validity" of the predictions in this book. This is not a scattershot exercise; all of the points in this book are carefully thought through, drawn from intensive research and scenario projects that I and others at Global Business Network have conducted, as well as on cutting-edge sources of information around the world. All of them are inevitable—that is, I have phrased and developed them to distinguish those that are certain to take place from those that are merely likely.

Some of these issues will probably be familiar to you, but others will be new. Some of the issues that you think are most obvious, and not even worth repeating, will turn out to be the most critical ones for other readers. For instance, there is a story about the human population slowdown in this book, which is so familiar to demographers that it has become second nature to them. They think it is obvious; that everyone already knows about it. But when I get up to talk about it, I find I can't simply mention it in passing. People practically jump up to ask, "But what about the exploding population?" I always have to take the time to explain, "It isn't exploding anymore." They may have heard this casually somewhere, but they've never absorbed the meaning of it. And some of the specifics will probably be thoroughly unfamiliar. I'd guess that, for most readers, this includes the physical underpinnings of potential teleportation described in Chapter 7.

Underneath the specifics, between the lines on every page in this book, you will find a basic message about the future in general: The challenges facing civilization right now are immense—arguably more difficult than they have been during the lifetime of any living person. At the same time, because of advances in knowledge and technology, the human race has never been so capable. And since most of our challenges are caused, at least partly, by our own activity, this expanded capability is a double-edged sword.

I am hardly the first to make these points. Indeed, for the past thirty years, ever since the publication of Alvin Toffler's book *Future Shock*, they have become part of the conventional wisdom. And yet most of us, in our decision making, are still not acting as if we truly believe them. The greatest challenge before us—on the personal, organizational, and societal levels—is to master our own accelerating power, without being swept away by it.

This does not necessarily mean taking hurried dramatic action. The dot-com bubble has shown the downside of rapid movement. The trick in navigating any set of rapids is being well prepared for all types of movement, and aware of the ways in which movement itself is changing. That's not an easy task, for rapids change all the time. The bottom of the river may only shift slowly, but the water level changes with the season and the weather. The rapids in the late spring can be a raging torrent of white water, while in the late autumn they can vanish as the water level drops. Rafting in the spring requires great skill and courage. The thrills are intense but the risks of falling out and being swept away by the current are great. Rafting in the autumn requires persistence in a slowly meandering river. The risk is running aground. The ride may seem less thrilling, and the water stagnant; but there is the steady satisfaction of endurance and balance. Navigating the future means being prepared to act in any season, and to shift from the mindset of one season to another as the environment changes. It means learning to recognize the rhythms of change before us, to avoid denial about them, and to practice our responses to them before they are upon us.

If you feel overwhelmed by the potential of surprises, your view of the future will be dark. You will continually expect to be hit by one unanticipated crisis after another, and your expectations will be met. On the other hand, you can approach the future with a sense of lighthearted enthusiasm—thoughtful, but eager to see what is coming next.

More than twenty years ago I was spending a Saturday afternoon at the Tassajara Zen Center not far from Monterey, California. It was a gorgeous day, and several of us were making our way down a slope clustered with pools, waterfalls, and rocks. The surfaces were very slippery, and most of us moved gingerly from one rock to the next, sliding along them, sitting on our bottoms when we felt it was too risky to stand. Suddenly a young woman came gliding down the hill. She looked like a dancer—slipping from one rock to another effortlessly and gracefully, perfectly balanced, never stopping for a moment. She had been walking those rocks so long, she felt comfortable on them. And it made the rest of us realize, suddenly, that the rocks were not such a dangerous environment at all. Not if you knew them.

The image of her movement has stayed with me, because I've seen people in many similar situations since then. We are terrified of movement because we don't know the environment very well. And then someone comes along who has thought about it, who is prepared, and who has a kind of uncanny grace and ability to move through peril. And even find time to be delighted by the challenge.

The Inevitable Surprises of 1978

This book looks ahead twenty-five years—the timespan of a generation of people. Before we get started, it's natural to look back twenty-five years and calibrate our judgment a bit. I had been in the scenario planning game for six years in 1978; working first at SRI International (formerly Stanford Research Institute), and then at Royal Dutch/Shell's famous Group Planning Department (where Pierre Wack and other colleagues developed the scenario-planning method, still in use today, that I discussed in my book *The Art of the Long View*).

What could we have seen in 1978? What were the inevitable surprises, visible then, that shaped our world of today?

- **Oil as a commodity:** Although we were still in the midst of the energy crisis, we knew that oil prices would have to fall. The industry was moving much too rapidly toward a multiple-sourced, flexible, trading-oriented model. And growing energy efficiency was driving down demand. OPEC would not be able to keep the price boosted artificially high forever. And, in fact, it didn't; the price crashed in 1986.

- **The end of the Cold War.** It was obvious to anyone who looked past the ideology of either communism or anticommunism that the Soviet Union could not much longer afford the expense of keeping its massive police-state empire intact. We didn't know exactly how it would end, but we knew it was unsustainable.

- **Radical transformations in the world of communications.** Fax machines, modems, electronic mail, satellite communications, cellular phones, and rudimentary precursors to the Internet already existed. The first wave of enthusiasm for personal computers had just begun, and some of the early applications (like computer-based spreadsheet programs) had appeared. It was clear that the results would be immense change in human communications capability and information gathering—as immense as the changes in mobility and infrastructure engendered by the automobile starting a hundred years before.

- **Nuclear power would vanish as a meaningful energy option.** The costs and risks were apparent.

- **Japan would go into a boom and then a decline.** The short-term protective benefits and ultimate systemic risks of their *keiretsu*-based financial structure (a structure of interlocking ownership and crony capitalism) were apparent.

- **The U.S. would be hit by a big violent-crime wave in the 1980s.** This was demographically driven; as the population of young men rose, so would crime.

- **The U.S. Savings and Loan Crisis** (or something like it). Deregulation in the U.S. and UK was a clearly unstoppable trend. Whenever there is massive deregulation of a previously highly regulated industry, there is a generally a crisis as the new institutions, without much memory of the old crises they engendered before regulation, try to test their limits.

- **The growth of the Asian tigers and the pressure on China to change course after the death of Mao Tse-tung in 1975.** It was not yet clear how China would move, because not much was known about the Chinese Communist party. But it would have to move somehow, because it, too, faced the same innate structural pressures that the Soviet Union did.

- **The rise of radical political Islam.** We were about to see the Iranian Revolution, which deposed the shah of Iran and created a Muslim state under the Ayatollah Khomeini.

If we look deeply enough, can we see the counterparts to these changes that are coming over the next quarter century?

CHAPTER 2

A World Integrated with Elders

In 2001 the United States reached a historic turning point that almost went unnoticed. The average age at which Americans retire hit a floor. It fell from 64 to 63, continuing a trend that had been steady for the previous fifty years . . . and then it began to rise again. In 2001 it was 64. In 2002 it was 66. And it is predetermined to continue to increase—probably as an accelerating rate. In the coming decades Americans, and others around the world, will retire on average at 67, 68, 69, . . . and on into their seventies and eighties. Within fifty years significant number of people will never retire; they will work productively through their deaths at ages above 100.

The graying of society is by now a familiar story in much of the developed world. There are ever-larger numbers of senior citizens in absolute terms; the numbers are even more significant when elders are tracked as a proportion of the population. Politicians cater to them at election time; they represent a perennial constituency of avid voters. A series of specialized businesses have developed over the last twenty-five years that cater directly to them, from retirement homes to private health care to travel and resort businesses. And through their bequests, trusts, and gifts they are one of the most important sources of capital and philanthropy in the world.

And yet, for all of that, people over 65 are still largely isolated from mainstream society, and ignored by most of it. The population at large is used to thinking of its elders as people who have checked out of the productive workforce, who live separately and have different priorities, and who have drifted into a postemployment, peripheral role in society, with dramatically diminishing capabilities as the years go by, and relatively little contact with the rest of us.

All of that is about to change. During the next three decades older people will become far more integrated into the rest of our culture than they have been since World War II. This shift has already started to take place; it's already a factor, I suspect, in the daily lives of most of the readers of this book.

The causes of this phenomenon can be traced to three distinct inevitabilities, each with its own inherent surprise:

First, the human life span is set to increase.

Second, the health of older people is dramatically improving, to a degree that finally realizes humanity's dream of retarding the aging process.

Third, the economics of aging is under immense pressure. This includes, of course, the familiar political pressures facing social security, health care, and pension laws. But on closer inspection, especially in light of the first two inevitabilities, the pressures will play out in a very different way than most policy-makers expect.

I. Life Span: Maximum Aspiration

Today, the oldest living human being is about 120 years old. This represents a leap forward in maximum life span, which has been rising steadily since the turn of the century. Average life span has also been rising. In 1950 in the United States, the average age at death was a bit higher than 60; today, it is around 77. A diagram of this trend, like the one in Figure 1 (on page 31) shows an average annual increase of about .67 percent, beginning at the turn of the twentieth century.

If you are reading this book within the first few years of its publication, then you are a direct beneficiary of this trend. No matter how old you are, your generation will, on average, have lived 5–10 percent longer, at least, than the generation of your parents. As the trend continues to unfold through the first half of this century, people will begin to routinely remain alive through their nineties and hundreds, at least in the U.S., Europe, and Japan. Many diseases that either are fatal today or that drain vitality and render people susceptible to other fatal illnesses will be eliminated, close to eliminated, or dramatically contained. This may include cancer in many forms; Alzheimer's disease and other brain diseases; diabetes; cerebral palsy; multiple sclerosis; heart disease; and a number of infectious diseases.

Does this increase in life span represent a one-time leap forward, a transition to a new stasis—in which most people expect to live as long as 100–120 years, but no longer? Or have we entered a time of perpetually increasing life span, in which people's expected maximum age continues to edge up to 130, 140, 150, and beyond?

Many reputable scientists apparently believe the former; or are, at least, making a case for a natural cap on human longevity of about 120 years. The magazine *Scientific American*, for example, recently assembled a group of fifty-one biologists and physicians to discredit the notion of a perpetually increasing lifespan. They see it as a form of hype, and a spur to spurious products like Human Growth Hormone or antioxidant supplements, which they implicitly portray as the twenty-first century's equivalent of patent medicines and snake oil.

Their report said that the "unprecedented increase in human life expectancy at birth," as they called it, could be traced to several factors: environmental technologies (such as sanitation and running water); medical advances (the use of penicillin, sulfa drugs, and antibiotics); new forms of health insurance, like Medicaid and Medicare, as well as their counterparts in other countries (this has the effect, among other things, of reducing suicide rates among the elderly); and the decline of cigarette smoking. That was the good news. The bad news: "Repeating this feat during the lifetimes of people alive today is unlikely." None of these critical factors, according to *Scientific American*, had any genuine effect on the fundamental limits to human lifespan per se. They merely removed some of the circumstantial impediments which kept people from reaching our maximum natural lifespan. Nor did these scientists foresee any further quantum leaps, either in technology or lifestyle, that could push the outer aging limit of human beings past that figure.

When you read their statement carefully, however, it is clear that the editors of *Scientific American* have hedged their bets on the issue. "We enthusiastically support research in genetic engineering, stem cells, geriatric medicine, and therapeutic pharmaceuticals," they wrote, "technologies that promise to revolutionize medicine as we know it. Most biogerontologists believe that our rapidly expanding scientific knowledge holds the promise that means may eventually be discovered to slow the rate of aging."

I personally believe that scientific research *will* succeed in extending the human life span, sometime within the next fifty years. We are at the threshold of a change in the nature of aging itself, and we can take confidence in the continued pushing back of aging, even though the details of the technologies have not always been developed, because of the body of theory that has been experimentally confirmed.

Two avenues of research are particularly suggestive. First, it is well established that severe caloric restriction has life-extending effects among many mammals. Rabbits and mice that consume minimal diets tend to live much longer than their conventionally fed counterparts. The same appears to be true for humans. If most people dropped their caloric intake by 30 percent, from around twenty-five hundred calories per day to around seventeen hundred, as long as they continued to eat nutritious food, they could delay their aging considerably. Very few people would choose to eat so little; we seem to be biologically hardwired to crave more food, probably from millennia of natural selection in which it was beneficial to build up stores of fat to burn during recurring times of scarcity. But by studying the physiological effects of low-caloric intake, we may be able to mimick them. Researchers Mark A. Lane, Donald K. Ingram, and George S. Roth have focused on the cellular metabolism of the sugar glucose.[2] Could it be, they ask, that lower caloric levels spur cells to develop more slowly and take extra care to preserve themselves? Could it be that, as Thomas Kirkwood of the University of Newcastle suggests, organisms balance the need to procreate against the need to maintain the body? We don't know for sure—but the fact that questions like this are being asked about a relatively well-established set of observable data means that it is reasonably plausible for some answer to emerge in the near future.

The second avenue of research, derived from cloning, has accelerated in the last five years. Aging is, after all, a form of damage to human cells, and cloning is now being used to regenerate and replace cells and tissues. One technique involves a fourteen-day treatment to generate new embryolike stem cells from the cells of aging people, with the *telomere*—a gene that one researcher describes as the "clock of cell aging"—manipulated with enzymes to a more youthful setting.[3] These cells could be used to generate new organs that might be transplanted into aging bodies, or they could be "seeded" through the body as replacements for cells that have worn out. Either directly, or by provoking other unexpected discoveries, these R-and-D efforts could push out the cap on human life span.

It's hard to say how quickly this will happen. It might take a hundred years to develop the necessary medical advances, or we might see startling breakthroughs as early as, say, 2007. It is certainly plausible that a significant number of human beings might live to age 150 by the year 2125. More likely, we will find ourselves able (with some medical interventions) to live to 120, and our children may live to 150. Even if none of this comes to pass, however, it is inevitable that a significant number of people in industrialized countries will live to 100. And that in itself is unprecedented in modern history. A society in which millions of people routinely live past 100 or 110 would be significantly different from any society that human beings have ever previously known.

In the industrialized nations of the world the effects have already been felt. Japan, the U.S., and Western Europe are all rapidly aging societies. In the U.S., demographers at the Social Security Administration expect the number of elderly people to double by 2035. Currently, there are five Americans in their prime working years (ages 20 to 64) for every individual over 65. Assuming no massive changes in immigration policy, that ratio will change to three-to-one in 2025. By 2075 the two groups will be virtually even in size.

The effects might even be broader. In parts of the developing world today, even in places like Africa that have been devastated by war, AIDS, and famine, people who aren't directly affected by those scourges are living longer. The biological phenomenon of life extension may turn out to be a form of epidemic—virtuous but contagious. It spreads from society to society, abetted in part by the spread of modern sanitation and medical technologies, but primarily by the budding awareness of healthful life practices. This awareness is building throughout the developing nations as people gain exposure to knowledge outside of their immediate communities. Even in the developing world, while the proportion of elderly people is growing smaller, the absolute numbers of aging people continue to grow and dwarf those of the industrialized world.

2. Aging: Raising the Stakes of Healthfulness

How old is that stranger across the room who appears to be 40? These days he or she might well be 50, 55, or even 60. A few years from now it may be hard to distinguish a 40-year-old from a wealthy 80- or even 90-year-old individual. The physical deterioration of aging is about to be dramatically slowed down, and in some ways even reversed.

As members of the baby-boom generation—those born between 1948 and 1962—become elders, they will tend to be far more noticeably younger looking, younger feeling, healthier, and more active than any of their predecessors. People will be vigorous and alert long past the boundary of what was once considered old age. Elders in the future—including many of the people reading this book—will work, travel, read, enjoy a full sexual life, remain athletic, and possibly even raise young children in their sixties, seventies, eighties, and nineties, with far fewer of the infirmities of aging than people at similar ages today.

The reasons for this are technological. The development of new instruments, the mapping of the human genome, the increasing research into nanotechnology, and the evolution of biogenetic and pharmacological research in general have all combined to accelerate each other. As any investor in biotech companies knows, biomedical research often fails to pan out; thus, we can't assume that any research under way in laboratories will be inevitably successful. For example: In 2002 a genetically engineered compound called imatinib mesylate, which is sold as the prescription leukemia drug Gleevec by the Swiss pharmaceutical company Novartis, turned 10 percent of patients' hair from gray back to their original hair color in a set of French field trials. Novartis immediately announced publicly that they were not researching Gleevec as an antiaging medicine. But they (or others) have a very great incentive to isolate the hair-color-restoring factor and produce it for the public. My own hair, which is now gray, was once a strong natural red. Will I get that pigment back in five years or so? It's hard to imagine that I could.[4]

But you might have said the same, fifteen years ago, if you'd asked if I could have the eyesight of a 7-year-old again. Or you might have argued that drugs to boost memory would always be placebos. Yet in 2002 a company called Memory Pharmaceuticals, cofounded by Nobel laureate (in medicine) Eric Kandel, announced six new drugs for the treatment of severe memory illnesses like Alzheimer's disease and dementia. These drugs, some of which stimulate enzymes that affect neuron effectiveness in various parts of the brain linked to memory, are also likely to improve short-term memory functioning for the broader aging population. Skin quality, bone strength, hearing, muscle tone, resistance to disease, and sexual potency—all have already been shown to be enhanceable through drugs, laser treatments, or other means. The drugs and treatments that enhance them will become increasingly sophisticated, powerful, and popular. At the same time almost all the degenerative diseases of aging—arthritis, osteoporosis, and various autoimmune diseases—will have been eliminated.

In his *Red Mars/Green Mars/Blue Mars* trilogy, science fiction writer Kim Stanley Robinson posited that "whole-body" treatments to reverse cellular aging could emerge by the end of the twenty-first century. In fact, the first such treatments for aging have already been administered, with mice as subjects.[5] These research efforts use genetically engineered viruses to bind new genes into a cell's DNA.

Much is unknown about this: Can the treatment be applied to human beings? What will be the side effects? And how successful will it be? But this is only one avenue of research, and within the next twenty years such announcements will proliferate. We will see more drugs, both internal and external, aimed at stopping cell decay through genetic intervention. Cyborg implants—machine augmentation of human physiological capability—will move out of the realm of catastrophic surgery and prosthetics, and prolong our everyday quality of life. Cochlear implants to deal with hearing loss are becoming common. Growing new glands to replace defective ones may be the next generation implant. Human habitats will become more and more oriented toward the reinforcement of longevity; research into cloning will ultimately yield cloned cells or implants that rejuvenate our existing physical selves.

Each of these innovations, in itself, will have relatively small effect. Some will fail. Others, while technically successful, will simply represent too much effort or expense for too little payoff (like Rogaine, which never found the vast market of bald men predicted for it). Still others will make a major difference in fighting particular diseases (Alzheimer's, heart disease, and strokes are three likely candidates) but may only represent small breakthroughs in preventing aging. The significant surprise will come from putting them together. They will all reinforce one another's effects, and the healthy way of life that becomes increasingly possible for more people will accelerate the rejuvenation effect.

What I've described so far is inevitable; the treatments exist today in some form, or are so numerous and well developed that it would take a miracle—or a very strong political movement—to stop them. And while such a political movement is possible, emerging out of religious and social concerns, everything we have seen so far suggests that protests would focus on very narrow issues, such as cloning, which would only partially defer the inevitability of the prevention of aging.

But there are also some uncertainties. The most significant has to do with the efficacy of the treatments themselves. Will they make aging reversible, so that 70-year-olds find themselves looking and feeling more like 30-year-olds? Or will they only be useful for those who are already youthful, allowing 30-year-olds to retain their physical presence while their parents are stuck with the physicality of 60? We also don't know how expensive such treatments will be. They may be rare and limited to a very few, or as commonplace as aspirin, affecting billions of people around the world.

Finally, it's not clear how widespread this trend will be geographically. We don't know, for example, how many political systems will be willing or able to pay the immense costs of developing rejuvenation treatments through national health services in the countries that have them. Already some countries with advanced medical facilities are becoming known as rejuvenation resorts, drawing a growing crowd of people who go there to be made younger. One of the surprising early indicators is the degree to which anti-aging treatments, as they are coming to be called, are popular in the developing world. Unilever executives, marketing the drug Retin-A (Vitamin A) as an ingredient in wrinkle cream, were initially surprised to discover high demand not just in the industrialized world, but in China, India, and Africa.

Even at a minimum, however, we're about to experience a dramatic boom in the capacity to lead productive and full lives after age 60. Many of the readers of this book will live to be 120; many of them will stay fairly youthful until at least age 100. You will look like 40- or 50-year-olds; you will be able to practice athletics, work, read, travel, and enjoy sex. You will lead full lives, even physical lives, unbound (as people have rarely been unbound before) by the infirmities of aging.

This, in turn, will fundamentally change the political and economic institutions around you.

3. The Institutions of Retirement

The political debates over retirement in the early twenty-first century all have to do with fear. We see a host of elderly people coming down the pike, with their public pensions underfunded, their private pensions uncertain, and no clear answer to the question of how we can ever pay for it. Missing from the debate is the most urgent question, the question of hope: "What kind of retirements will people *want* and *choose*?" Inevitably, those retirements will look and feel very different from the retirements that most people in the United States and other wealthy countries are planning for today.

Two people I know epitomize the change for me. The first is an 85-year-old woman—let's call her Grace—who lives in a retirement community outside San Francisco. Her social calendar is busier than those of most of the middle-aged people I know. She is vigorous, athletic, and urbane in a way once associated only with the wealthiest and most vigorous aristocrats. Although she does not work for a living, she is active in several local philanthropies and other organizations, and she is often asked for advice by businesspeople who live near her.

There are hundreds of thousands of people like Grace in America, more and more every year. They often have pensions—theirs, their spouses', or both—that give them $100,000 or more per annum to spend. They have adult children who have long since launched careers of their own, so they have no dependents and few responsibilities except their own nest egg to manage. They are proficient on bicycles and skis, fashion conscious, and intellectually curious. And they look and act as if they have many more years of vigorous life ahead of them. Among their ranks are a growing number of well-known people who have continued to conduct research, publish work, and manage others long past retirement age: economists like Milton Friedman, historians like Alfred Chandler and Barbara Tuchman, management thinkers like Peter Drucker, and many, many more.

But I also think of Sarah, a flight attendant whom I met recently on an airplane. She was clearly in her early seventies, struggling to find the strength and stamina that her job required. I asked her why she continued to work. "I can't afford to retire," she told me. "I'm not married; I don't get paid that much. And I barely have retirement benefits." Once, she would have been forced out by a mandatory retirement age. Now, mandatory retirement is seen as discriminatory, so her fate and the airlines' are locked together—until the airline is forced into layoffs or she becomes too ill to continue working. She could easily have another ten years on the job. Sarah, too, is typical of many people who will be growing old in the years to come.

In the wealthy nations of the world we face nothing less than a thorough redefinition of retirement. To be sure, people will continue to retire from their jobs. But retirement itself will be a far less dramatic transition. It will take place later in life, or with less of a distinct break from ordinary worklife, and with the general assumption that people will resume work in some form—at least after their first three or four careers. Retirement will no longer be the twilight of life, focused on rest, recreation, and recuperation; it will now be a time in which people will marshal their experience and intelligence to create entirely new lives for themselves.

It's obvious why this will happen. When the projected life span was 70–75 years, it was natural to retire at age 65 with the expectation of a few "well-earned" years of vacation. But if we can now expect to be vigorous through age 110 or 120, then it makes no sense to shuffle off to a retirement community for fifty-five or so years of collecting checks. To be sure, there will also be macroeconomic pressure to raise the threshold retirement age for government-sponsored benefits, so that the pension systems of industrial countries can still afford to pay them. But an equally strong pressure will come from retirees themselves: it will be boredom. Wealthy and active people like Grace won't want to stop working. More impoverished people, like Sarah, won't be able to stop working, even with their social security checks as a cushion. In both cases the money from social security will be inadequate and, for many, practically irrelevant.

In a sense we are about to return to the policies and practices of the early twentieth century, when there was no such thing as "retirement age," because very few people lived that long. When social security payments were first instituted in the United States in 1935, they were seen as a safety-net measure that would only be needed by most people for the few years remaining to them after retirement (and like child labor laws, they were also a way to free up jobs for working-age people). There were 80-, 90- and 100-year-olds (the first recipient of a social security check lived to be 100), but they were comparatively rare.

It was only in the 1950s that the average American's life expectancy expanded to the point where a lifestyle focused around retirement could emerge. By the mid-1970s the "golden agers" and "snowbirds" who ran (for example) the Florida-Maine and Texas-Colorado circuits had become a subculture in their own right, with individual life expectancies of twenty years or more after they left their jobs. They lived longer than they ever had before (at least in aggregate); they had healthier lives after retirement than most people had had before; and their work-free income from pensions and social security continued through the end of their lives, supported by the payments of younger workers into the system.

In that context the shift in the direction of retirement age—the historic moment in which average ages of retirement stopped falling and started rising, back up to age 66 and 67—is an indicator of a deeper set of changes that have already begun to take place. Starting now, and for the foreseeable future, average life span in America will rise faster than retirement age. As the chart below shows, this has already begun.

Average U.S. Life Spans and Social Security Age

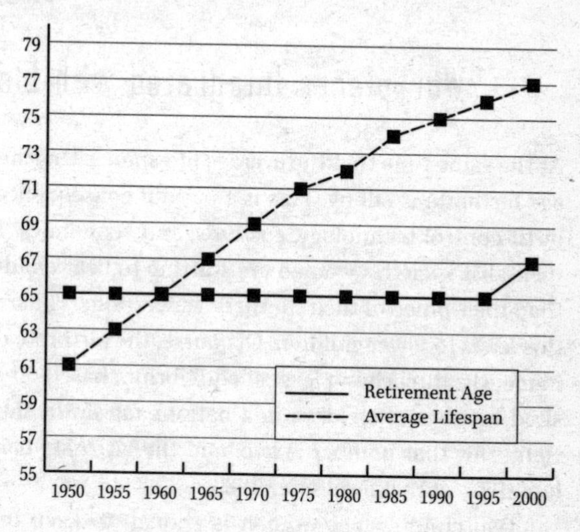

This graph shows the average age of retirement compared to the average life span in the United States over the past fifty years. Note that the gap between them has been widening since the early 1960s; only recently have people begun to push back the age of retirement, for reasons either of financial security or personal satisfaction.

And that demographic shift has also been echoed in official policy, at least in the United States, with a prescient set of Social Security Law amendments signed by President Ronald Reagan in 1983. Beginning in 2004 the legal age of retirement—the age of eligibility for social security benefits—will increase to 66. This means that people born in 1938 or later will have an extra year of work before they can collect benefits. In 2035 it will rise another year: people born in 1959 will have to wait until they turn 67 to collect benefits. The progression of the law, after that, is uncertain. If there is an outcry of protest in 2004, then that will probably slow down the change. But the pressure to keep pushing the retirement age higher will persist.

And that, in turn, inevitably changes a great deal about the workplace and our society.

Workplaces Integrated with Elders

At the same time that the average life-span is lengthening, the average birthrate is falling. This is a natural consequence of changes in birth-control technology, customs, and economics: Even in fundamentalist societies women are starting to bear children later in life than the women of their mothers' generations. For any given mother this leads to fewer children. Of course, the birthrate doesn't fall permanently; it reaches a level of equilibrium based on the average desired number of children in a nation's family. In the United States right now that number is two, and the birthrate has been more or less static at that level since 1980.

Two children per woman is enough to keep providing young

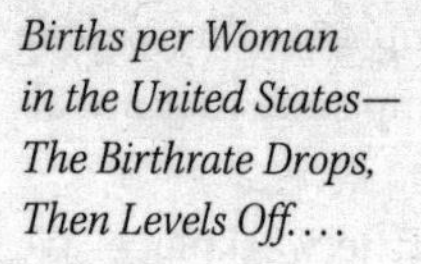

Births per Woman in the United States—The Birthrate Drops, Then Levels Off. . . .

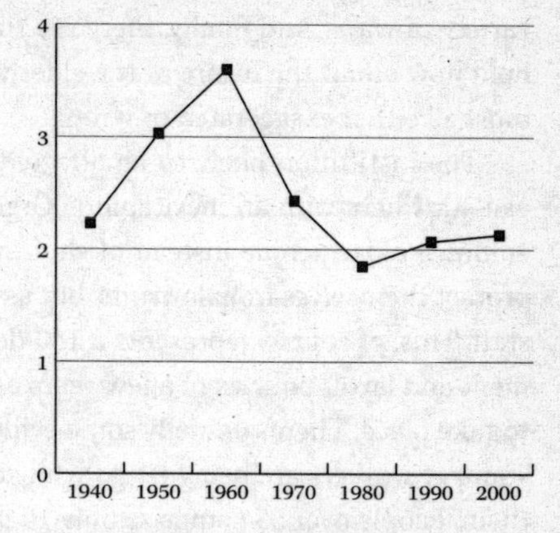

people in society, but no longer enough to dominate it, as it was dominated by children and youth during the "baby-boom" years of the 1960s (in the U.S.) or the "global teenager" years of the 1980s and 1990s (around the world). Where once the "population bomb," as Stanford professor Paul Ehrlich called it, was seen as an explosion of babies, it is now going to be an explosion of the elderly.

This is not a linear change, but a logarithmic one—the kind of change during which it seems for a long time as if nothing much is happening; then, suddenly, the effect shows up as an enormous surprise. This particular inevitability, for the most part, will come as no surprise: writers like Ken Dychtwald, Peter Peterson, and Theodore Roszak have published popular books on the "age wave" that is about to engulf society, and the proportion of older people is much larger than it has ever been in human history. The demographic "pyramid" that determines the source of political and social culture is clearly about to change radically. And yet, both in policy and culture, people have barely begun to recognize the ramifications.

To make sense of this we need to separate three sets of concepts. First, there are the inevitabilities: the aspects of this shift that are certain to take place, because they have been set in motion already. (At the very least it would take a dramatic shift of some sort to stop them.) Second, there are the critical uncertainties: the factors that will make an enormous difference, that could play out in a variety of ways. And finally, there are the assumptions that people hold now about the future of the elderly—assumptions that are almost all either exaggerated or wrong.

The institution likely to be affected first is the workplace. We can start here with an inevitability: Organizations will increasingly embrace older people instead of shunting them away—not just to protect themselves from lawsuits, but as a means to a more effective staff. This, of course, represents a 180-degree turn from the recruitment and layoff policies of a few years ago, but it has already begun to take place. There's actually some evidence that, for the first time, younger workers are being let go while older workers are replacing them. People over 55 comprise only 10 percent of the workforce today, but they account for 22 percent of the job growth in the United States since 1995.[6] Even during the recession year of 2002 the labor-force participation rate for older workers (aged 55–64) jumped two percentage points. Andrew Eschtruth, the associate director of the Center for Retirement Research at Boston College, called this increase "unprecedented in postwar U.S. economic history."[7] A study of 232 U.S. employers found that 60 percent of them had some policy for rehiring retired people; General Electric's "Golden Opportunity" program, for example, allows retirees to work up to a thousand hours a year.[8]

In the past there has been a widespread reluctance to hire older people because they were about to retire soon. They didn't have much future; why invest in them? But if an individual feels and acts like a healthy 40-year-old through the age of 95, then he or she could be hired at 60 and work twenty-five years or more, with vitality to spare. Already, people over 55 are staying on the job for an average of fifteen years after being hired. That figure will only increase. In such a world many people will never retire; they'll simply go on working until they become infirm. They will change careers, reeducate themselves, and continue to learn and produce as they go.

Many employers have also been influenced by the assumption, either explicit or implicit, that older people are less productive than younger people. But studies comparing productivity between older and younger workers find either no difference, or even a slight edge for older workers. A lot of the difference has to do with perceptions, by others or by the elder employees themselves. If people believe that "an old dog can't learn new tricks," then they don't. But if they believe they can adapt, they do. And they have other advantages. Older people need less training in general. They are absent less often for short-term sicknesses or family emergencies; even at age 70 they use fewer health-care benefits than employees with young children.[9] There is evidence that, despite some impairments like memory loss, they are capable of coping with far greater levels of complexity.[10] (And with the growing effectiveness of pharmaceuticals I mentioned before, coupled with the continuing proliferation of information technologies, memory loss won't be nearly as much of a problem a decade from now as it is today.) Most importantly, they are more effective in enterprises that depend less on human stamina or coordination, and more on judgment and awareness. Even mass-manufacturing assembly lines and mining operations require less brawn and more brain; since technology changes so rapidly, the seasoned and multifaceted judgment of elderly people carries more of a premium.

Many older managers or knowledge workers today will be hired for their third, fourth, fifth, or sixth career tomorrow, and consequently bring a cross-disciplinary "hybrid vigor" perspective that will be increasingly valuable. Many of them are immensely flexible; adults over 50 constitute the fastest-growing group of Internet users. They are not like older workers of past generations, who typically stayed in one company for a full career. A 55-year-old in 2002 has been employed, typically, in three or four organizations since 1970, with a variety of responsibilities in a world that has changed dramatically. Such people are predisposed to be flexible and to internalize the lessons of the mistakes of their past. And as Theodore Roszak points out, they have an understanding, tempered by a lifetime of experience, that recognizes "wisdom, compassion, and the survival of the gentlest." All of this comes together when they take a new position.

I've recently become a venture capitalist and, because of this, sit on the board of several fledgling companies. Like many experienced VCs I am useful to the extent that I can make vibrant and relevant the lessons of my own experience as a corporate executive. For example, I made a number of mistakes when I cofounded Global Business Network; mistakes I would like to think I'd avoid if I tried it a second time around. As an advisor to other companies I can help keep them from repeating those errors. That's typical, these days, of the amplification effect as older people build partnerships with younger people in a knowledge-based society.

Rich and Poor Elders

As employment changes, so will the infrastructure that currently supports retirement. This, too, is inevitable. Legal retirement age may remain at 67, but conventionally acceptable retirement age will tend to disappear. "Social security" will no longer be seen as retirement income; it will be seen as a stipend given to people above the age of (perhaps) 70 or 75 as a small but politically shrewd acknowledgment of the debt that society owes them. Even today, social security income is too small to live on for most retired people; researchers Richard Burkhauser, Kenneth Couch, and John Phillips have found that the single most significant factor in determining whether people retired or not was whether they had a separate pension. Early takers of social security are increasingly using it not to support themselves, but to springboard a new business.[11] That trend will continue.

Whether or not social security is privatized, in short, turns out to be uncertain—but relatively unimportant. Other factors are more critical in determining the quality of life for this newly elderly society. Chief among them is the crisis in health care: an uncertainty, but one that is less dire than many professionals in the field assume. To set the stage for this, let me note the very prevalent assumption that we are entering a severe "haves-and-have-nots" society, in which the gap between rich and poor will continue to increase. Some elderly people (like Sarah, the flight attendant) keep working long past retirement age simply because they have no choice.

This assumption has some validity. Because wealthy people will be living longer to invest and reinvest their money before it's divided among their heirs and the government after their death, it is indeed inevitable that the disparity between "haves" and "have-nots" will widen over the next twenty-five years, at least in some critical respects. The gap will be particularly pronounced among the aging. Sarah, for example, will be more productive and vigorous than she would have been two decades ago. She will live a decade or so longer than she might have lived, had she been the same age in the 1970s. But she will never be able to retire. She will have to keep working until she is too exhausted and worn down to continue. She is too poor to quit. Sarah's life will either be tolerable or dismal as she ages. The major distinguishing factor is health care costs. Can she afford the cocktail of drugs and treatments that would rejuvenate her enough to make a longer work life tolerable?

Currently, health care costs are rising (in the United States at least), at a rate of about 15 percent per year. Thus, a family of four paying $10,000 per year for health insurance in 2002 would find that bill increasing to $26,000 or more by 2010. Such rates, of course, are unsustainable for political and economic reasons—and they won't be sustained. Something, inevitably, has to give. Otherwise, people will be mortgaging their houses to pay for catastrophic health care because they can't afford the insurance. In a democracy, that would be politically unviable. We'd vote out of office any politicians who had permitted it to happen on their watch.

In fact, the system is reformable. We know this because we have been here before. In the 1970s, and again in the 1980s there were big run-ups in health care costs. Both times the system got under control by restructuring—the establishment, for example, of HMOs and other forms of capitated insurance. We still manage and use our health care resources very unwisely; there is still an excessively burdensome system of administration and cost avoidance, and an ineffective, inefficient system of allocating health care resources. Thus, there is an enormous amount of room for improving the way in which the system works, reducing costs, and improving quality and service. When the pressure for reform grows high enough (and it isn't there yet), these inefficiencies will be resolved.

Why haven't they been resolved already? Largely because of taboos that stem from perception and fear of lawsuit. The most effective way to cut medical costs is to stop investing so much money in the last thirty days of a patient's life. I can speak to this personally. My mother died of cancer not long ago. There is no question that she consumed 25–40 percent of the health care costs of her entire lifetime during the last six weeks of treatment. She was hospitalized, and kept on a variety of devices designed to keep her alive as long as possible. It was, in the end, a futile effort and a heartbreaking one. It added unnecessary tension, uncertainty, pain, and frustration to her passing. And for each individual who is kept alive that way for a few extra weeks, the entire system loses money and other resources that would allow it to care more effectively for all its elders as a whole.

To change this we need a change of attitude. Today, doctors are compelled, by custom and law, to override the interests of their patients and keep them alive heroically, even at the expense of the patient's quality of life. Changing this state of affairs would have been unthinkable in our twentieth-century era, when health deteriorates in peoples' seventies and eighties, before they and their children are ready. But in a future world of healthy 90-, 100- and 110-year-olds, it may be far more bearable to think of alternative approaches to death. More people may be willing to embrace the philosophy that says, "Yes, there comes a moment when we realize we are headed to the end. We can delay our passing somewhat, but not very long. So how can we manage the exit with grace, instead of keeping ourselves pumped up as long as possible?" In other words, more family members may be willing to move from the intensive care unit to the hospice as a place for the final days of their loved ones' lives.

Many elderly people already favor this for themselves. Hospice care is a model for compassionate, quality care at the end of life; it emphasizes the relief of pain, management of symptoms, and psychological and spiritual support. A typical stay might be twenty-five days; some are much shorter, and a few last for half a year or more. In 2000 about one out of four Americans who died did so either in a hospice, or while receiving similar care at home. Though cost is usually not the principal factor behind the choice, the cost impact is enormous; a hospice care receiver saves $3,000 or more over an intensive-care-unit patient during the last month of life. The number of people receiving such care has grown exponentially in the U.S.—from fewer than twenty-five thousand in 1982 to more than seven hundred thousand in the year 2000.[12]

That is just one example of the cost savings to be retrieved from the health care system. And the pressure to retrieve them, while not inevitably successful, will inevitably grow stronger. For the health care system is going to be asked to do many more things, that often will inflict extensive costs. Youthful vigor will not come cheap. It will sometimes require surgery; other times, newly patented drugs. It may sometimes entail new forms of treatment, like the many ingenious (and highly expensive) methods that now exist for boosting fertility for couples over age 40. Anything involving cloning or genetic manipulation is likely to be quite expensive.

Inevitably, therefore, there will be several tiers of care. One tier will serve working-class people like Sarah, the flight attendant, who will still live longer and be healthier than they have in the past, but who will be relative "have-nots." There will be another tier, continually expanding and growing more sophisticated, for the upper-middle class, the kind of people who (currently, in 2003) might save up for a year to pay for a fertility treatment. And then there will be a tier for the wealthy, who might routinely spend thousands of dollars per year on new capabilities (including the first genuine efforts to extend the life span past 120). Advances in each tier, of course, will rapidly trickle down to those below; just as laser eye surgery is now within reach of working-class salaries, so will some of the antiaging advances of the next two decades become widely available. Here, then, is the great uncertainty about health care. How many of its advances will be broadly affordable? We don't know. The more of a mass-market item longevity becomes, the better life will be for people like Sarah, who are forced to work.

Of course, neither Sarah nor Grace represents the extremes of wealth and poverty that we will see coming ahead. We probably can't imagine the full extent of wealth *or* poverty that we will see on either side. But we already know that, inevitably, there will be surprisingly large and unexpected populations at both extremes.

The poverty side will be augmented by an inevitable and yet largely unseen problem population that has, so far, existed below the surface. About 6.5 million prisoners are currently in state or federal prison, serving sentences of twenty-five years or longer for various felonies, often related to violence or illicit drugs. Most of them have been arrested since the mid-1980s; many are poorly educated, from impoverished backgrounds, and unprepared for life outside prison. Starting around 2010 they will be released in unprecedented numbers. Some cities around America will be inundated with aging former felons.

The United States doesn't have a very good record with this population—either before prison, during prison, or after prison. Demographically, they are a question mark. We might expect them to return to crime, except that crime is a young man's game. Most crimes are committed by men between ages 18 and 35. These men will be muscular and tough in some ways, but they will also have aging bodies and no easy access to income. Most of them are ineligible even for social security benefits.[13] Their work records are spotty. They have no prospects. What are they going to do? Imagine that you are such an individual. It's 2015, and you're 50 years old, and you've been locked up for thirty years for breaking and entering and shooting someone during a robbery. How will you survive?

This is a significant inevitable surprise. We know it will happen. But we don't know what consequence it will have. One can imagine, for example, a new wave of AIDS; in 1997, one quarter of the people living with AIDS or HIV in the United States had come out of prison that year. One can also imagine a great boost in the population of street gangs and crime syndicates, often organized around racial lines: the Aryan Brotherhood, the "Mexican Mafia," the Bloods and Crips, and the old Italian crime syndicates. These entities, which deal in prostitution, narcotics, and protection rackets, may themselves be rejuvenated by a massive new population of members. Or they may find it impossible to support their new population.

If this inevitable surprise gives you shivers, then you may find it helpful to think about an equally inevitable, equally surprising development at the other end of society: an enormous intergenerational wealth transfer over the next twenty years. The generation currently in its seventies and eighties will hand over something like $10 trillion to its children. Along with that will come an immense boom in philanthropy. Katherine Fulton writes in her report, *The Coming Flood*:

> If the trends of the last quarter century continue, [philanthropic] giving in this decade will be half a trillion dollars more than in the last. But when you add in the intergenerational wealth transfer that is already under way, the numbers go up substantially, adding another one-half to one trillion dollars. In other words, by 2010, an additional $100–150 billion will likely be given away in the U.S. each year. Similar forces are at play in Europe.

At first glance this might seem difficult to believe. People will be living much longer; why would they have money to spare? Because an extra thirty to forty years of health and work can make an enormous difference to a portfolio. The longer that hardworking professional people live, the more likely they are to accrue wealth, rather than to dissipate it. There will be millions of people over 55, with children grown and past college, at the peak of their earning power, with forty or fifty years ahead of them during which to pay off their mortgages, accumulate savings, build up real-estate portfolios, or accumulate wealth through other means. Not all of them will use that time wisely, but many will, if only because they have learned from the bitter experiences and risks of their earlier years.

Those who never stop working will have more than enough money for expenses—even for the expenses of longevity treatments—and no great incentive to give all of it to their children (who are similarly set up to build wealth portfolios of their own). Even if the tax incentives for giving bequests and the recent move to curb estate taxes do not hold, they will be moved to set up family trusts simply as a way to contribute. Compared to the Rockefeller, Ford, or Gates foundations, these family trusts will be minuscule; but in aggregate they will represent an amazing wave of philanthropy. And there will also be an unprecedented amount of innovative capital, available from "angel" investors who have grown wealthy enough to fund a few new risky enterprises of their own.

A New Way of Thinking about Values and Lifestyles

At least one inevitability will *not* be much of a surprise: the continued growth of businesses that cater to elderly people. These include spas, cruises, Club Med–style villages for older people, high-tech outposts, and other types of high-end vacation environments. Homes that combine medically assisted living with active recreation, including skiing and bicycling. Recreation-oriented states like Colorado, which bulged with children in the 1970s, young adults in the 1980s and 1990s, and middle-aged adults in the past few years, will soon become dominated by people over 50. We will see booming businesses in amusement parks for the elderly, in clothes for older customers, in cars with amenities for older drivers (like radar to detect objects ahead), and in cosmetics for "mature skin." According to Stefan Theil of *Newsweek*, the cosmetic brand Shiseido's "Benefiance" and Lancôme's "Absolue," both marketed to older people, account for half of those companies' skin-care sales.[14]

There's a very high probability that our fundamental expectations about the pattern of life will begin to alter as a consequence. Education, marriage, work, will be mixed in different combinations as life extends farther and farther and farther into the future. We've already seen people getting married later, forming families later, having babies later. The basic pattern of life will be made new, just as radically as it was when we moved from the beginning of the last century, where the life span was between 35 and 45, to a world today in which we can expect to live to more than 100.

More surprising perhaps, but equally inevitable, is a growing shift of attitude about marriage and family: multiple generational families, multiple sequential families, multiple careers. On the one hand, in the future we will see many more multigenerational families. But those who live very long are less likely to remain mated for a lifetime. A married couple in 2002 who are both in their seventies, with children and grandchildren, have little reason to divorce. They're used to each other, and they expect to live for only a few more years. But what if they looked ahead to another thirty-five or forty years? Would they still feel as committed to spending the rest of their lives together? In many cases the answer would probably be no. Adultery will still be unacceptable. But serial monogamy will be the norm. So the surprise will be a return to higher divorce rates.

The result will be complex, multigenerational, intersecting families of a new sort, featuring stepparents, stepgrandparents, siblings with age gaps of fifty years or more, and internecine patterns of interlocking responsibility. There have been multigenerational extended families in the past, particularly among the wealthy, but these will represent a very different kind from any we have seen before. Many will be multiracial or multiethnic; they will involve people who, at one point or another, dramatically changed their lifestyle or moved from one part of the world to another. They will involve new forms of inheritance, new living arrangements, new kinds of communities, and ties that could span six generations in a single family.

We do not know the difference that all of this will make to our political and social values. In the past people leaned toward being more conservative as they grew older, in part because they became more accustomed to an established way of looking at the world. Now we will have a massive older generation with unprecedented wealth and health—people used to continuous learning, who have reinvented and uprooted themselves several times during a lifetime. They will dominate many institutions, probably including most political institutions. When Senator Strom Thurmond reached his hundredth birthday and retired in the same week, he provided an example of longevity that many others in the future will emulate. No doubt a whole section of many legislatures will be set aside for wheelchair access.

More likely than not the widespread employment of older people will become a great engine of productivity in the countries that encourage it: Japan, China, the U.S., and parts of Europe in particular. Of course, it could turn out that the elderly, while they may be vigorous in a physical sense, become fearful and unwilling to change. Japan's elderly have created such a society today, and their will, imposed on the rest of their country, is one reason for Japan's extended economic slump.

While I can't prove it, I have to believe that most of the elderly people of the future will be more adventurous and experimental. While writing this book, I happened to spend some time with a close friend who had made a lot of money by age 55. But he had never quite figured out what to do with that money. He was financially independent and miserable. He was uninterested in religion, politics, or starting a new business. Golf was boring. Flying his private plane had ceased to fascinate him. What was he going to do with his life? He is looking desperately for an answer.

One part of him wants to be adventurous and try something completely new. Another part wants to hold on to his dignity and comfort. This is not a bad kind of tension to have among a lot of capable, experienced people. It means that ideas and experiments can get tested in ways that only a few individuals would have had the perspective to explore in the past.

James Hilton expressed this kind of perspective very well in his novel *Lost Horizon*, a novel about a hidden community of people who had discovered the secret of longevity. The leader of the community, at one point, tells the narrator of the book: "You are still, I should say, a youngish man by the world's standards; your life, as people say, lies ahead of you. In the normal course you might expect twenty or thirty years of only slightly and gradually diminishing activity. By no means a cheerless prospect, and I can hardly expect you to see it as I do—as a slender, breathless, and far too frantic interlude. The first quarter century of your life was doubtless lived under the cloud of being too young for things, while the last quarter century would normally be shadowed by the still darker cloud of being too old for them; and between those two clouds, what small and narrow sunlight illumines a human lifetime!"

Now, a startlingly large number of us will be stepping out of that narrow band of sunlight into a great, extended day.

CHAPTER 3

The Great Flood of People

The trickle of migrating humanity, which has been gathering momentum since the mid-1800s, is finally taking its permanent form: a continuing, ongoing flood.

It used to be possible to dam the flood of humanity—to keep people from migrating from one country to another. That's not possible anymore, at least not in the United States or Europe, and probably not in Asia, Africa, or Latin America either. In this age of multiple vehicles, powerful transborder communications, and permeable boundaries, you can't close off countries anymore to immigrants. If they can't come in through the immigration services, enough of them will find a way. (More precisely, they will find a way whenever there is a land route. Australia can still close its borders to immigrants, and probably always will be able to. Water routes in general can be blocked; relatively few boats make it from Morocco to Spain, from Albania to Italy, or even from Haiti or Cuba to Florida. But whenever there is overland passage, the flow is essentially unstoppable.)

Traditionally, most people who migrate are fleeing instability and persecution or looking for opportunities that they could not find in their homeland. They are often destitute when they arrive and, with a highly vulnerable legal status, are easy to exploit. They settle near other people from the same homeland, recognizing that, in this new, unfamiliar environment, with a fraction of the resources that are available to natives, they need help from each other. They can rarely find the work they seek; there are many lawyers and architects from Latin America or Pakistan now working as cabdrivers in New York or waiters in London. In continental Europe, where they are often barred from employment by law, many of them end up on the public dole (perhaps for generations—we don't know yet, because most are first- or second-generation immigrants, and it is unclear what will happen to them over time). On the other hand, because they often bring youthful energy and ambition with them into aging societies, they tend to be harbingers of economic vitality.

Many people reading this book are descended from migrants—certainly, most of the readers in North and South America, and many of the readers in Europe as well. I myself am a first generation American; my parents were Hungarian Holocaust survivors (my mother in Auschwitz and my father as a slave laborer) and I was born in a refugee camp in Germany. With this history in our minds and (in a sense) in our bones, it's natural for us to think of migration as an arduous passage: a test for the people who make the trip, and a test for the society that accepts them. It is also a test of the society that they left, of course. But when the tide of migration becomes a flood, as it is today, then it represents a fourth kind of test: a test of our theories of opportunity.

Do you believe in the theory of unlimited potential: that wealth creates more of itself, and can grow indefinitely? If so, then implicitly you are in favor of migration. People coming into a country will help it grow.

Or do you believe that there is a limited quantity of wealth in a society—that there will always be some people on the top and some on the bottom, and the proportions (while they might shift a little) will essentially remain the same? If so, then you will necessarily see migration as an evil, because for every immigrant family that betters itself, an established family will be forced farther down the social ladder.

In a sense, migration has put both these theories to the test and has proven both of them. In conditions where people expect that wealth can grow indefinitely, migration becomes a vibrant linchpin of society. On the other hand, in societies where wealth is seen as a stagnant, finite quantity, migration becomes one of the most stressful forces in existence.

This distinction between America and Europe will inevitably become even more marked. As a culture believing in unlimited opportunity and potential (also known literally as "the American Dream"), with policies that make it relatively easy for people to rise above their original wealth and education levels, the United States will continue to be a thriving destination for ambitious immigrants. My own personal experience, like that of many readers of this book, is an example of the facility with which the right kind of society can help people realize that dream. My parents never went to college. They were always relatively poor. When they migrated to the United States, that meant I had the opportunity to go to college, despite their lack of wealth. If you were to rank Americans by net worth today, I'd be in the top 2 percent. But I am not exactly unique; there are another 4 million people in the U.S. at least as wealthy as I am and that number will inevitably grow—both in absolute numbers and in proportion to the whole.

Europe, by contrast, is a culture steeped in the belief that people cannot easily change the destiny they are born into. Many of its politicians, consciously or not, embrace ingrained policies and practices that are designed to preserve the wealth and status that people already have. (The U.S. has such policies and practices, too, but not nearly to the degree that Europe does.) But there is also a compensating factor. Precisely because its culture does not believe in the alternative of individual progress, Europe takes far better care than we do of its "have-nots." The result: Europe's economy is much less dynamic, and its migration problem is far more devastating. Solving the migration problem may be so difficult that it tears the European Union apart. Conversely, if Europe successfully weathers the storm of migration during the next few years, it can probably weather anything.

The migration story, therefore, is a story not of politics or demographics but of cultural challenge. Do the people of the most powerful regions in the world—North America, East Asia, and Europe—have the intellectual and emotional wherewithal to absorb and foster the floods of humanity coming their way? We don't know the answer yet. But we know the floods are coming, and that a great deal hangs in the balance.

China: In Search of Wives and Work

One inevitability is no longer much of a surprise: the economic resurgence of China and its impact on the rest of the world. Once, businesspeople around the world yearned for China's markets to open. Now they are afraid of Chinese industrial competition. Bolstered not only by an enormous population of cheap laborers, but by changes in its own government's policies and a sudden wave of technological innovation and foreign investment, China is becoming an economic superpower. Its economic growth rate of 6–7 percent per year dwarfs that of most other nations (only the other sudden Asian powerhouse, India, has a similar growth rate). But as the well-known Japanese strategy writer Kenichi Ohmae pointed out in his recent study of the Chinese boom, that growth rate is only the tip of the iceberg. Region-states along the coast, such as Shenzhen, Shanghai, and Dalian, are now growing at 15–20 percent per year. This is far faster than the Asian "tigers" of Malaysia, Taiwan, Thailand, and South Korea.[15]

The rapid growth rate of Chinese manufacturing can be easily understood as a consequence of globalization. But it is also driven by the need to provide goods and services to the Chinese people themselves: a market of millions of people who can afford appliances and automobiles (for example) for the first time in their families' history. China, in short, has the most dramatic and potentially influential emerging middle class in the world.

"To find a precedent for a society comparable to China in 2002," writes Ohmae, "you would have to go back forty years, to the Japan of the 1960s, which was diligently preparing itself to become a global competitor. There are also echoes of Dickensian England—the dawn of the industrial revolution—and America's 'robber-baron' era of the late 1800s, when the United States first showed signs of becoming a global economic power."

Ohmae and other commentators (most notably Orville Schell, who is a longtime writer on China and currently dean of the School of Journalism at the University of California at Berkeley) have noted that the continued growth of China as an economic power is not *necessarily* inevitable. There are still major "internal contradictions" (as Schell puts it), including institutional corruption (among the country's financial markets, for example), environmental degradation, the free flow of information over the Internet (vital to the booming economy but deadly to one-party rule if unchecked), and the "crisis of belief" that is developing as people start to leave behind the ideology of communism.

However that plays out, two demographic trends in the future of

China *are* predetermined. Both represent the unintended consequences of policies put in place twenty years ago, just after the death of Mao Tse-tung and the rise to power of Deng Xiaoping. Both are fairly surprising—that is, they are both transforming the status-quo relationship between China and the rest of the world, in ways that few leaders are prepared for, particularly in China itself.

The first was the famous "one family, one child" policy. Both incentives and punishments were put in place that made it very difficult for Chinese parents to raise more than one child. But the officials who enforced the law didn't specify anything about gender, and Chinese parents took matters into their own hands. Using ultrasound to detect the sex of their unborn fetuses, they tended to favor males and abort females. In 1990, for every 100 girls born in China, 111 boys were born—a difference of three or four percentage points. By 1995 there were 116 boys born for every 100 girls—a ratio that continues today.[16]

Five percentage points may not seem like much, but when there are over 10 million people born every year, this translates into 1/2 million "excess" males coming into the population each year for the next twenty years. (It could be much longer than twenty years, unless China reverses the policy or unless female children suddenly become culturally more desirable.) This trend is especially significant in a Confucian culture like China's, where a high premium is placed on marriage. There are literally millions of Chinese men with no one to marry. They are therefore spreading out into the rest of the world, in an unprecedented find-a-bride-style diaspora—seeking degrees and training along with matrimony as they go.

Naturally, they are going to cities where there already are large Chinese populations. Singapore, San Francisco, Vancouver, New York, Lima, and London are popular destinations. They stay in the homes of cousins or family friends who have emigrated from China in the past, and who may have introductions ready to young women in the neighborhood (unless all the eligible females are already monopolized by other male Chinese). Typically, however, this is a short-term diaspora. Since the governments of Chinese provincial "region-states" (which have a great deal of local power, particularly in the arenas of quality of life and education) are actively courting young Chinese professionals to move back to the mainland with their newly acquired professional and business skills (and since they can probably make more money in China's newly entrepreneurial economy), it's likely that their stay outside their homeland will be brief.

That doesn't mean the migration is insignificant. In fact, it has two noteworthy effects, both of which will make China more interdependent with the outside world than it has ever been in modern times. First, Chinese men will go out into the world in large numbers, seeking wives and education, for anywhere between one and four years. Then they will return to their country. This is unprecedented. Since the Communist party took over in 1948, Chinese people have left, but very few have returned.

Second, it suggests that there will be an immense increase in immigration of non-Chinese people *into* China, despite its current crowdedness and the fact that it is still a totalitarian country. Chinese men, as members of the new middle class of a booming economy, are in a position to attract women from (for example) Pakistan, the Philippines, Southeast Asia, and even India. The size of this flow depends on how open the Chinese government becomes to immigrants from other cultures, particularly those who are coming to make a life for themselves and raise children.

"Note that the increasing skewed ratios over time mean that shopping for younger women within China will *not* be an option," remarks GBN demographer Chris Ertel. "If I were forced to speculate, I would say that there will be some strong pressure for sex-specific proimmigration policies around 2010 in China."

After fifty years of relative isolation this new migration will establish a new pattern of interdependence between China and the rest of the world. It will increase the family ties that already exist, and thus provide a natural channel for flows of investment in both directions. It will almost certainly lead to far more travel back and forth between China and other nations, not just by businesspeople, but by ordinary people visiting relatives. And it will further shore up China's position (as commentators like Ohmae see it) as the "new America"—a new land of opportunity for people wishing to improve their economic position.

Which brings us to another huge Chinese pattern of migration—this time within the country. Millions of people are moving from the farmlands of China's interior to the burgeoning new industrial communities on the coast. They are going where the economy is prospering. And they are doing it fast. It took the United States more than a century to go from a society where 10 percent of the population lived in metropolitan areas (in 1835), to a society where 70 percent of the population did (in 1960).[17] In all that time the population was never greater than 180 million. China has 1.2 billion people, and they're going to make a similar transition in the space of a few years.

"This is the largest migration in human history," comments Schell. "One hundred fifty million people are migrating from the countryside into the cities to pick up work with no welfare, no schools, no health care, no houses, nothing. In cities like Beijing, Canton, and Shanghai, roughly thirty to forty percent of the populace are what they call 'floating population.' This works to China's advantage in certain economic ways, namely by providing low-cost labor to build freeways, tunnels, buildings, infrastructure. But it works to China's disadvantage if there's an economic slowdown, because suddenly you then have tens of millions of displaced people. There is no context for them. They can't go back to the land."[18]

The business leaders of the Chinese coastal cities are, of course, delighted. Kenichi Ohmae interviewed the manager of an electronic components manufacturing plant with ten thousand workers, all young women, all earning about $80 per month. When Ohmae asked why none of the women wore eyeglasses, the manager told him that he fired them as soon as their eyesight went bad. "They can find another job—that's not my problem. There are plenty of people who want to work for us."[19]

At the same time the central government is apparently scared silly. They've mounted campaigns to try to keep rural people on the farm as long as they can, to stem the influx into the cities. But they can't—not for very long. Because at the same time, they're doing all they can to increase the agricultural productivity of the land. Maoist farms tended to be small, divided plots, on which one person grew enough rice to feed three. Those days are disappearing, and there isn't enough for farmers to do. They can either head to the cities or be a burden on their relatives.

Nor is there much room for them to head into. China's population problem is literally centuries old. Every habitable space has been occupied for as long as people can remember. There isn't anything like the virgin forests of North America; the population is larger than Europe's and even denser. And in that sense the current wave of migration may represent the saving factor for China's future.

If you want to understand the choice facing China today, visit Bangkok and then visit Singapore. If the new cities of coastal China come to resemble Bangkok—a place where people flock with little infrastructure to receive them, little productive work, no provision for electricity or running water, and live on squatters' land on which only ramshackle dwellings are permitted to be built—then the new society is probably unsustainable. China may, in fact, have reached the limit of the carrying capacity for its people.

But if China's new urban centers can come to resemble a hundred Singapores—a hundred dense but brilliantly designed and managed cities in which efficiency and quality of life are placed at a premium, and civic society is managed like a well-run corporation—then its population probably has plenty of room to thrive in. The governors of China's local provinces are emulating Singapore; indeed, Singaporeans are training them. But it's still too early to say whether the emulation will succeed.

In any case, it will not be a smooth, well-coordinated process. There will be fits and starts: godawful conditions, immense crowding, and then a slum clearance and modern construction on the old site. Earthquakes won't be seen as disasters, but as opportunities to clear some of the rubble away and put down new infrastructure.

Some places will be virtually uninhabitable by contemporary standards. On the other hand, Pittsburgh in 1949 was very close to uninhabitable, and today it is considered one of the most livable cities in North America.

We can get hints of the potential for the new China by looking at some of the building sites where the migrants from inland are settling. Beijing, for example, has responded to this enormous migration by changing its energy source. Only ten years ago, most homes in China's capital city were heated by small coal braziers, burning brown coal and releasing it through a smokestack in the living room ceiling. In its ecological impact this was one step cleaner than burning dung, and the air quality of Beijing was famous for its coal-dust soot. Today, most of those braziers have been (or are being) replaced by natural gas—a much cleaner fuel. The air is, accordingly, already much cleaner—or it would be, except that China's ubiquitous bicycles are rapidly being replaced by automobiles and diesel buses.

One urban development, called Pudong, has been constructed on the marshes and islands of the Huangpu River delta outside Shanghai. It isn't precisely a city; technically, it is still governed by Shanghai. It is more like an immense multiskyscraper industrial project, but with housing and recreational facilities as well. Like Brasilia it was designed and conceived as one elaborate piece of work. Unlike Brasilia it seems designed to be resilient and responsive to the people within it; it may not end up as dreadful and sterile as Brasilia, but instead, it may stand as the prototype for the urban environments of the Chinese future, platforms of human habitation in which the new middle class settles into vertical, rather than horizontal, suburbs. Hong Kong is already an example—this sort of urban design allows an efficient concentration of population with business, manufacturing, and mass transportation along with accessible green spaces. If China can pull off this transformation, it won't just solve its own habitat problems; it will make itself an exemplar to the rest of the world of the kinds of innovations needed to squeeze so many moving people into such crowded urban spaces.

The United States: Cultural Complexity

If you want to know the demographic future of the United States, including the bedrock, Midwestern, "pure" United States, study California. More than 45 percent of the incoming freshman class at the University of California at Berkeley in 2002 are of Asian descent. Hispanics (people who can trace their ancestry to Latin America or the Caribbean) represent about 15 percent. Whites—people with European ethnic backgrounds—represent only 30 percent. Together, in other words, Asians and Latinos outnumber "Anglos" by a factor of two to one.

When the numbers were released, there was an almost audible collective gasp—even in Berkeley. Only a few years ago the Asian-American incoming class would have been no more than 15–20 percent. Suddenly, in the space of one year it had tripled.

On one level this was a consequence of the end of affirmative action—a university policy designed to eliminate racial or ethnic preferences, and to admit people more purely on the basis of academic and character merit. One result: Black Americans comprised only 3 percent of the incoming class, down from 10 percent two years before. That decline may or may not have been expected, but I doubt that the opponents of affirmative action expected the corollary result: the enormous percentage of Asian-Americans who perform highly on tests and other academic scores, and who thus gain acceptance. (I don't know a convincing reason why they do so well, other than coming from family backgrounds that promote diligence and study. But even other diligent, study-oriented cultures do not produce the educational credentials that Asian-Americans amass.)

In 2025, when these students are in their mid-forties and early fifties, a large proportion of them will be leaders in the elite corridors of corporations, nonprofits, and government. Even if these elite jobs are disproportionately skewed toward whites in the future (as they no doubt will be), there simply won't be enough whites to go around. Nor will they always be the most desirable candidates. The ranks of power will be increasingly filled by those who are currently called "people of color"—particularly Asian-Americans and (to a lesser extent) Latin Americans. This will mark an enormous shift from the power elite of forty years ago, or even from that of today. It will resemble the dramatic increase in the number of women in the professions—especially law and medicine—over the last thirty years. It will ripple into the culture in ways that are difficult to define; but that will inevitably be significant. Americans are about to experience the demographic surprise of waking up one morning to discover we're not a white country anymore.

The great European migrations pretty much ended with World War II, except for an odd influx of Hungarians (in 1956) or Russians (in the early 1990s). In the last twenty-five years the United States has been flooded with newcomers from Mexico, Central America, and a few South American countries (particularly Chile). There have also been wave after wave of émigrés from Southeast Asia, particularly after the Vietnam and Cambodian conflicts. Millions of people have moved to the United States from Indonesia, the Philippines, Thailand, and India. And there has been an immense wave of Chinese immigrants, particularly in the last five years, which began as the United Kingdom prepared to honor its agreement to return Hong Kong to China, picking up momentum as the Chinese gender imbalance induced more men to emigrate. And it is still gathering today.

Both the Asian and Latin American immigrant populations of the last few years include some very well-educated and well-trained people. Most of the migrants will do almost anything to feed their families. They will be the taxicab drivers, hard laborers, gardeners, and mechanics of the next fifteen years. Their children, starting around 2010, will enter the leadership class of this country.

Drive up University Avenue in Berkeley or Atlantic Avenue in Brooklyn, and you can identify the communities from their restaurants. Currently, in Berkeley, you can find enclaves of Salvadorans, Pakistanis, Chileans, Thai, and Vietnamese. Already, the members of these communities are starting to climb up from their immigrant ghetto status to become active parts of American society. They are going to the great universities. They are entering the executive tracks of leading corporations. They are starting to occupy political office.

This doesn't mean that each of these groups is moving into positions of leadership en masse. There are a large number of Latin Americans who have low levels of education, and who will probably never break out of the lower middle class. But by 2015 at the latest, and perhaps much earlier, the elite of the nation as a whole—the leadership class with predominant influence in business, media, politics, the arts, education, and community life—will include people of a variety of ethnic backgrounds. This may not seem like much of a surprise in San Francisco, New York, Miami, or Chicago. But it represents a shift of perspective for, say, Des Moines, Spokane, Minneapolis, Houston, Wheeling, Biloxi, and Bangor—and for the suburbs between major cities. Most American communities already have some people from diverse backgrounds, but the local tables of power and privilege have reserved their seats for the children of old-line families. That is about to change. Already, African-Americans are sitting at or near the head of the table. (Two obvious examples are Colin Powell and Time Warner CEO Richard Parsons.) Soon, they will be joined by large numbers of Latin Americans and Asians.

The ascension of immigrants into positions of leadership and respectability tends to come as a surprise to each new generation. America was startled in the 1930s by the rise of people of Irish and German descent (whose great-grandparents had migrated here in the mid-1800s) to positions of eminence. It was startled in the 1960s and 1970s by the sudden prominence of second- and third-generation Jews and Italians. Each time, there is a reaction among the "old-line" political interests, who don't want to see their influence diluted. But as we have seen in recent years, the reactions never take hold. Republicans in California tried to campaign against immigration in the mid-1990s, and nearly destroyed their state party in the process. There were too many "people of color" and people of other diverse backgrounds already placed in too many positions of influence in the state.

The same is true of the rest of the country. Despite its reputation in some circles for bigotry, the U.S. is the most immigrant-friendly society on the planet. We have large immigrant communities that make it easy for newcomers to acclimate themselves and get an economic foothold. There is very little political will in the U.S. to reverse our acceptance of migration, because too many Americans are themselves the descendants of immigrants—and we recognize the degree to which the national vitality depends upon continued immigration. This, in turn, makes the United States more attractive to people outside our borders. In the twenty-first century America will continue to draw to itself more immigrants, per capita, than any other major nation.

Migration is always a fundamental driver of long-term societal change. Each group that assimilates into a culture subtly shifts the style of the culture as a whole. This is particularly true when they gain access to elite status, which is typically accomplished through education. Arguably, the last people to succeed in this way (before the current wave of Asians and Latin Americans) were the Jews. There had been significant numbers of Jews in the United States since the 1850s—first from Germany, then from Eastern Europe—but they were limited to particular professions, like retail merchandising and engineering, because most universities refused to admit more than a handful of them. Then in the 1950s, as the "Greatest Generation" returned to civilian life, the concept of merit took hold in the U.S.; it was no longer considered acceptable to discriminate against Jews. Suddenly, Jews who were performing at a high academic level zoomed up to positions of power and influence in investment banking, fashion, the arts, and even government. In the era of Robert Rubin, Ralph Lauren, Barbra Streisand, Steven Spielberg, Joseph Lieberman, Diane Feinstein, and Ari Fleischer, it is a bit startling to realize that all of these arenas were once considered, if not exactly closed to Jews, then a bit too exclusive for them.

Something similar is happening for non-European ethnic groups today. For that reason I believe the current debate over multiculturalism, passionate though it may be, is short lived. An Irish-American who marches in the St. Patrick's Day parade, or an Italian-American who marches in the Columbus Day parade, may put on ethnicity as a beloved garment from time to time; but he or she reserves the right to take it off at will as well. And the same will be true for other groups, once they assimilate to the point where they are ready to assume positions of leadership as Americans. Multiculturalism—the view that society should be organized as a set of separate ethnic cultures, each with its own political and social identity—is a naïve view. It might or might not be a worthy ideal, but it is unattainable. The real world, in the United States at least, is a much more tangled mix of origins, evolutions, and interrelationships. The boundaries between subcultures are already being bridged—even the once seemingly impassable boundary between the black and white races. This is symbolized by two things: first, by the ever-increasing rate of intermarriage across racial and ethnic lines, as described in Randall Kennedy's new book, *Interracial Intimacies: Sex, Marriage, Identity, and Adoption.* Second, by the ever-increasing appearance of Latin-American and Asian-American characters in popular television and motion pictures. It's significant that one of the main trio of characters in the recent movie *Charlie's Angels* was Lucy Liu, a Chinese-American woman. It's much more significant that most of the public took her ethnicity for granted; it was hardly mentioned in reviews of the film.

The most intriguing question about, say, the year 2025 is not how America will change its Latin, Asian, and other ethnic subcultures, but how they will change America. Imagine being a sophisticated observer of the cultural scene in the 1950s, watching Jews enter the universities in quantity for the first time. Would you have been able to predict the influence they would have on American culture? Clearly they weren't the only influence, but they were significant. Would you have imagined Americans growing more urbane and literate, more brash and funny, less oriented toward a "country gentleman" ethic, more skeptical of religious fundamentalism and mysticism, and more willing to tolerate (and argue with) different points of view openly, rather than sidestep them in hushed tones?

What, then, will be the values that Asians and Latin Americans bring to arenas like media, government, and business? The answer isn't simple, in part because the situation isn't simple. It isn't as if the "Asian values" simply get added as an ingredient to a dish called "American values." Nor are Asian values uniform: the cultures of Thailand, the Philippines, China, Korea, and India are all extremely different. Nor is it appropriate to equate "Asian-American" values with the values of people who never left their original country. Many of the Asians who enter elite universities, for example, will be third- or fourth-generation Americans. They may not speak their parents' or grandparents' languages. They are used to thinking and speaking in English. They aren't Asian any more than an American whose great-great-grandparents came from London is a "British-American."

But it is worth noting a few common cultural elements that Asians share—and that white Americans do not. The European Protestant culture has a deeply embedded ethic about progress. It is not just feasible, but a moral obligation, to better oneself, even if it means contradicting authority. Most Asian cultures see people as having an innate place in life that they cannot break out of; to try is to succumb to an illusion. Many Asian-Americans live with the continual tension between this Confucian view and the fact that they or their parents *did*, in fact, break out of their place in society, simply by virtue of coming to the United States. They may be entrepreneurs as individuals, but they do not have the same kind of entrepreneurial culture as a vehicle from which to launch themselves. Like other immigrants before them, the new Asians will struggle with old-world values in a new cultural context.

The U.S. will remain primarily an English-speaking country, though Spanish, in particular, will make inroads. Unlike most Asian and European languages, it is spoken by people from a wide variety of countries and backgrounds. One inevitable surprise will be most evident to linguists in, say, 2025—the degree to which Spanish phrases, terms, and ideas have migrated into conventional English, as it is heard both on the street and on television. Chinese, by contrast, will be a specialty language in America, like Italian or Yiddish. We may expect to see a book sometime in the next ten years which lampoons that fact: a book called, perhaps, *The Joy of Mandarin.* The publication of that book might serve as a signal to the rest of us that Asian-Americans had finally recognized that their culture was now a pivotal part of the American mix.

All of this may come as more of a surprise to the American white population, who will wake up one day and find that they aren't speaking "pure" English anymore; that, in fact, this isn't the country they thought they were in.

Just as it did for China, migration offers a test for the United States. Can we develop a truly multicultural society with no majority culture? Can we get past the kind of splintered society in which various "identity groups" manage their own part of the community? Can we create a society in which people from different ethnic backgrounds, racial backgrounds, and perspectives create a social milieu together without having to be the same? This is a novel test, and it is already under way; one can be optimistic about how America will meet it. The challenge may be more extreme today, but Protestant America successfully, if reluctantly and with a fair amount of violence, absorbed the Catholics of Italy, Poland, Ireland, and Greece along with the Jews.

Europe: A Strain on the Family

Since the 1960s Europe has been the target of a long wave of migration, primarily from the Muslim world. Millions of people have emigrated from Turkey, North Africa, the Caucasus, Afghanistan, Turkey, Pakistan, and India, to Germany, France, Britain, the Netherlands, Belgium, and Scandinavia. Millions more are continuing to come. Some are moving to find work; some as refugees; and some because they have relatives in Europe. In a very real sense this migration is the final chapter in the story of colonialism. Indians and

Pakistanis go to London and Manchester; Algerians to Paris; Turks to Germany. The colonial attitudes, which Europeans exported for centuries, are finally coming home to roost, long after the colonies themselves became self-governing.

Almost all of this migration has been, and continues to be, illegal. Europeans, who have never been welcoming to migrants in general, are powerless to stop this wave. However, it's taken them more than thirty years to understand this—enough time for the Muslim immigrants to build up, in effect, their own subcultures within the larger European nations. There would be hostility even if the migrants were blond, blue-eyed Christians, but since they are dark-skinned, Arabic-speaking Muslims, they are often treated with contempt. In cities all around Europe we see played out an enmity that goes back to the twelfth-century Crusades.

Many Muslims feel their Christian hosts are lazy, decadent, and exploitive. Why, then, do they come? Because the jobs are on the Continent. There are no jobs in Morocco, so Muslim youth cross over to Spain to work in car factories. (During the past thirty years Europe's governments exacerbated the problem by treating their would-be immigrants as "guest workers": employing them but denying them the dignity of an opportunity for citizenship or any political influence.)

By 2025 there will be very large Muslim communities in virtually every major European country. Since Muslim women tend to have children early in life (and often), these enclaves will have by far the highest population growth rates of any part of Europe. Some of the Muslim population will be dependent on welfare; they will be sources of crime and social conflict. Others may well be the bearers of new entrepreneurial energy and vitality; they could become active contributors to the economic well-being of their host societies. Rich or poor, however, they will all have one thing in common: They will all be set apart. There is little interest on either side in having them assimilate.

Even in the 1970s, when it was still a trickle, immigration from the Muslim world was seen as a social problem in much of continental Europe. Rainer Werner Fassbinder's film *Fear Eats the Soul*, about a Moroccan immigrant in Munich who marries a 60-year-old German charwoman, was controversial back in the mid-1970s. The trickle accelerated into a torrent after the Cold War ended, when routes opened up from former Soviet republics and satellites: Albania, Kazakhstan, Bulgaria, Romania, and Russia itself. The ongoing war between India and Pakistan induced a growing number of people from both countries to emigrate. The Gulf War widened the torrent further; Palestinians and Pakistanis who were thrown out of Kuwait and Saudi Arabia joined in. More recently, the massive problems of Africa have led to huge migrations across the Sahara and Mediterranean.

In the early 1990s the Maastricht agreement opened up the borders among European Common Market countries. Now anyone who is a resident in any European country can go anywhere in Europe. Newcomers make their way into Spain, Portugal, or Italy, which are relatively easy to enter from outside. Once resident there, they can take a train up to France, Germany, or the Netherlands without difficulty. No one, apparently, foresaw the consequences. London is now 15 percent Muslim, with the percentage of Pakistani and Bangladeshi girls taking A levels in 1998 more than double that of white boys that same year.[20]

France is the current flashpoint of Islamic immigration in Europe. It has the highest percentage and population of Muslims in Western Europe (7 percent and 4 million)[21], and the most complicated ex-colonial relationship with their former host countries. Germany has the second largest Muslim population (3 million), followed by the U.K. (1.8 million), which has a deeply historical intertwined relationship with its Muslim ex-colonies, and fifty-four hundred Muslim millionaires (measured by cash and stock but not property).[22] Close on the heels in absolute population of Muslims is the Netherlands (750,000) followed by Sweden and Switzerland (both with 350,000).[23]

All of these countries have developed very deep tensions. Observing from the United States, where immigration has been a great source of economic and social dynamism, we may be surprised to find the same phenomenon in Europe. But there are two critical differences. First, the caliber of immigrant is different. Those who are university educated or creative and entrepreneurial tend to choose the United States to emigrate to, if they have a choice. They perceive the opportunity to be greater there. If you are a Pakistani computer scientist, you'll be drawn to Silicon Valley to program supercomputers or launch your own company, not to a London suburb to operate a launderette. Chances are, if you had that level of education in Pakistan (or Algeria, or Morocco) you already emigrated long ago. Your less-educated cousins (who are also less interested in technological education or any work-related skills) are more likely to emigrate to Europe now.

Second, Europe has thus far been closed even to those migrants who demonstrate brilliance. I know an extremely capable Lebanese executive who has never been able to get past a midlevel position in his company in Paris. After all, he is not French. His kids go to good schools; he has a beautiful home and good health care. But he will never realize his potential as a leader and innovator. His children will never make it into the local elites, either, nor will their children, unless the society changes. To an American this is a tragic waste. But to my Lebanese friend it is simply part of the rules of the game that he accepted when he came there.

And that's the most positive situation imaginable for a migrant into France. For most of my friend's fellow immigrants the realities are far worse, and far more tragic. There are no jobs for the migrants into Europe. They are not welcome. But there they are, and they must be taken care of, or else they will become social problems. So they are placed on the dole—given minimal payments and health care. They settle into the poor quarters of Amsterdam, Paris, Marseille, and Munich. They do not speak the language in the same way, or come from the same culture, so their children do poorly in school. Relatively few are dependable workers. Many have children—the birthrate of Muslim migrants is far higher than that of their established European neighbors.[24] They take part in petty crime; there are crime syndicates of Muslim youth in Amsterdam, for instance, who specialize in distracting tourists on the trains and stealing their luggage. And there are gangsters in Lyon from the Côte d'Ivoire who shuffle back and forth between prisons and the streets.

Some migrants become candidates for terrorist cells. Others import or deal drugs. Many are violent. Arab youth in Antwerp held a two-day riot in November 2002, demanding political control over their neighborhoods and all-Arabic schools. The instigators apparently had ties to the Lebanese terrorist group Hezbollah. Rioters smashed the windows of stores that didn't display a pro-Arab sticker.[25] At the same time there have been violent episodes where native Germans, Belgians, French, or other Europeans taunt, chase, and violently attack Muslims.

All of this creates a terrible social and political dilemma for Europe as a whole. Since World War II the European nations have developed a deeply entrenched set of cultural values in favor of taking care of the weaker members of society. When you live in a small country (or a closely knit region such as you find in many parts of Italy, Germany and France), everyone you meet feels, in a sense, related to his or her compatriots. You accept the need to take care of each other because you are all part of one big family. Now, imagine that suddenly a group of poor relations, or (worse yet) strangers have forced themselves into your community home. You have to take care of them; that's your ethic. But you find their crowdedness and their habits revolting, you yearn for them to be gone, they are ungrateful for all you have done for them, and you wonder how many of them you can support.

The problem is predetermined to get worse. Can anyone preserve their old way of life in the face of this kind of pressure? The answer is obviously no. But Europeans have already gone through a great deal to come to their current set of values—their facility for peace, "soft power," high environmental quality, and civilized lifestyles. They are not going to abandon these lightly. Their ability to deal with this conflict, and work their way out of it, is the primary test they will face in the next two decades. And it is not at all clear that they will pass the test.

The resulting political tensions have only started to appear. The reactionary right is gaining ground in many European countries precisely because there appears to be no other group taking a stand against immigration. Some are unabashed anti-Muslims or anti-migrants: National Front party presidential candidate Jean-Marie Le Pen in France, the assassinated politician Pim Fortuyn in the Netherlands, the Danish politician Pia Kjersgaard, Bavarian governor Edmund Stoiber, Jörg Haider in Austria, and many more.[26] Their numbers, and their popularity, will continue to grow. On one level a stand against migrants does little good—you could just as easily take a stand against a hurricane—though it feels, to many, like the only vehicle for response. On another level such a stand raises old horrific associations with fascist nationalism, and the implications are hard to avoid.

Europeans will increasingly be polarized between those who prefer the idea of a more "multicultural society" and those who want to regulate immigration tightly. It's not certain which side will win, but there is a very good chance that, whichever side wins, "immigration police" will be developed to fight crime and terrorism in Muslim communities. Europe, the bastion of social democracy, may be the place where universal ID cards are first established by governments—as a way of keeping tabs on their migrant populations. We will see more right-wing governments come to power in Europe. Some will be former social democrats who have changed their political views; others, like Le Pen and Stoiber, will be long-standing champions of chauvinistic nationalism. Different countries may take alternative measures, but no country will be *too* tolerant. The reward for that would be having all the migrants from various nearby countries flood into your borders.

If the European Union were the United States, we would expect this migration to follow a cycle of assimilation, with each new generation of immigrants developing its own path to full integration into Europe's economy and culture. But Europe has no such history except ancient invasions from the East, and its wave of immigrants is too new to have developed a precedent. We don't know what will happen as a result.

At worst this flood of immigrants could tear the continent apart. One can imagine a scenario in which Muslim terrorists with "guest worker" visas could blow up the Eiffel Tower, the Reichstag, and Big Ben all on one day. Instead of triggering a unified response, such an event could fragment the European Union. Each nation could react by closing its borders. Internal security conflicts could tear down the trust developed among governments. New border controls lead to reinstated tariffs. The Euro would crash and countries revert to their original currencies. Within thirty years war could once again ravage the continent.

But there is also a plausible optimistic scenario. This migration could revitalize the continent. It turns out that there *is* intrinsic merit in diversity, and there is untapped scientific or engineering brilliance in the ambitious and hardworking people pouring into Europe. Everything depends on the economic conditions in which the immigrants find themselves. If the economy is robust; if there are technologies and capital available; if people are willing to invest in creating a developmental path, that can employ immigrants effectively and lead them into more productive work; then society tends to be resilient enough to handle the stress. The 1950s was a great time to emigrate to the United States, precisely because the information age was taking off, and there were many jobs available with developing career tracks. My father was able to come to the U.S. and get a job as an electronic engineer, and he spent the next fifteen years learning to apply his old knowledge in a major new field.

Suppose, then, that you are a European leader today—in government or business—and you see this stress coming. You don't know when, but you know it will hit a breaking point eventually if nothing is done to change direction. What would you do?

First, avoid denial. Plan for the shift, and don't fight it. Don't harbor the illusion that border patrols can cut the flow of humanity by ninety percent. Accept that it will be at least several million per year.

Second, don't try to exhort the existing population to accept the new arrivals. You will never convince them. Instead, raise the quality of the new population. Invest enormously in education for them. It takes twenty years to educate a citizen, and that may not seem like a quick enough return on your investment, but there aren't any short-term solutions to this problem. The only tactic left to you is to do whatever you can to change the game fundamentally in the long run. Education—both adult and childhood education—is the single biggest lever you have.

Third, rethink your penal and welfare policy in light of the new realities. Redesign the legal frameworks to recognize the key distinction within the immigrant community—between potential productive people and people who will never be productive. Use that distinction to decide which people to support and which to deport.

Finally, the best way to deal with the immigration problem is to make people want to stay home in the first place. Which means stimulating economic growth, prosperity, and freedom in the former colonial countries from which emigrants are coming. You can best help these failing countries by helping them become more successful. Incidentally, this is the official U.S. policy toward Mexico, and it appears—since the advent of NAFTA in 1994, at least—to be having an effect.

Given what we know about Europe's values and culture, this is, of these four measures, probably the one that gives it the best chance of handling this inevitable surprise.

CHAPTER 4

The Return of the Long Boom

When the dot-com stock bubble burst in April 2000, the public mood was sanguine at first. In many minds this was just a mild correction. Investors believed their portfolios would be back to the booming growth rates soon.

But things haven't gone back to normal. Large numbers of middle-class people have quietly watched their mutual funds and retirement funds lose half their value or more in the past three years. Just in the first month of 2003, 94 percent of the stocks in the

Standard & Poor's 500 Index declined; major averages went down about 12 percent. People who once saw themselves retiring in the Bahamas at age 55 are now wondering whether they'll have to work the rest of their lives. The rest of the economy has also been affected. Employment levels, while not falling dramatically, haven't risen much either. State governments find themselves with enormous deficits. Enron, WorldCom, and other corporate scandals have demonstrated that the business practices of large corporations are not always trustworthy. Fears about the consequences of the war on terrorism have underlined the relationship between market bullishness and global stability. Japan and much of Latin America are enmeshed in local economic crises, and there is concern the United States may soon join them.

The prevailing conventional wisdom is increasingly anxious. It says that we're in for a long, deep, and dismal economic downturn, a "double-dip" recession with perhaps a return to the kind of deflation that Japan has struggled with for a decade, and that the United States last saw in the Great Depression of the 1930s. In this context there is an economic surprise coming that may be hard to swallow, but it *is* inevitable: This downturn won't last. The Long Boom is coming back.

That doesn't mean that the current economic malaise (and there is no other word so apt for it) will be reversed very quickly. The global economy will probably remain stalled for another two or three years, for reasons that I will make clear later in this chapter. Short-term gains or losses are *not* inevitable; it's hard to know what the economy will do in the short term, because there are too many intermingled players with short-term influence and constantly shifting short-term goals.

But the long-term trend is still inevitable. The underlying factors that created the boom in the first place are still in place, and still evident. They never went away. If you spend some time, as I have, watching these forces at play, you will see why the "Long Boom" economic scenario of the mid-1990s—in which we enter a great global economic expansion that allows literally billions of people to move into middle-class lifestyles around the world—is still on track to occur.

As it happens, I had a great deal to do with the popularization of the term *the Long Boom.* First in a 1997 article in *Wired* magazine, and then in a 1999 book with Peter Leyden and Joel Hyatt, I used that phrase to propose the idea that the world could be entering into an unprecedented period of prosperity. The *Wired* article, in particular, appeared at a point when young people coming out of college perceived themselves as entering an economy that had been mired in doldrums for twenty years. The glittering eighties had been viewed by the popular media as a sham: lots of froth for the Gordon Gekkos at the speculative top, but very little real wealth creation for the Archie Bunkers in the middle. The idea that the world could move into a period of sustained prosperity was extremely hard for people to swallow, and I couldn't blame them for their skepticism.

Looking back, I prefer the skepticism of the mid-nineties to the evangelical fervor for the new economy that followed. During the next few years the Dow Jones Industrial Average rose to 12,000, and it was easy to conflate the stock market bubble of the late 1990s with the Long Boom. But they were never the same thing, and my colleagues and I were always very careful to distinguish them. In fact, to anyone attuned to the Long Boom, the market bubble was frightening; it couldn't last. (I remember waking up one morning in October 1999, while on a trip to Spain, and calling my investment manager to say, "Take me out of everything that's in the stock market." I put it all into bond funds. I missed some of the upside that way, but I couldn't bear the tension any longer.) Whereas the Dow rose and fell dramatically during the five years between 1996 and 2001, the Long Boom trend has been much steadier—a linear progression of about 5 percent growth per year beginning in 1983, with what we believe will be a forty-year cycle in its course. If you look at the trends since 1985, you can see that the Dow, despite its ups and downs, has never fallen below the Long Boom line.

I couldn't fault anyone for being skeptical about the return of the Long Boom today, after the wild ride we've been through. But the idea of the Long Boom does not come out of wishful thinking. It was originally derived from observing the effect of two fundamental forces:

First, economic productivity is genuinely increasing. There really *is* a new economy, evolving out of computer technology and innovative management, allowing business (and ordinary people) to deliver unprecedented levels of accomplishment with greater efficiency and faster returns for every investment of time or money. Industries around the world are steadily improving their quality and reducing their costs; moreover, a series of new industries are poised to take off, and some old ones (such as the automobile industry) are on the brink of transformation. And productivity works: it is probably the most consistent vehicle we have for enlarging the pool of human wealth in the world.

Second, globalization also works. Controversial though it may be (and as we'll see, there are good reasons for some of the controversy), the integration of trade and employment among nations turns out to be a genuinely effective way to broaden the base of worldwide prosperity.

There was a third force that we didn't fully recognize in the mid-1990s, but that has now come into full relief in the light of the bubble-bursting of the 1990s:

The innate value of transforming infrastructure. The existing grids of telecommunications, energy, traffic, water, finance, and education take on lives of their own, and some of them last a very long time. The roads in some cities follow paths laid down a thousand or more years ago. No wonder it takes three hours to get from one end of London to the other. The Long Boom accelerates each time there is an effective transformation of an outmoded old infrastructure. That kind of transformation has been one of the quiet but pervasive characteristics of our time: just consider the impact of the credit card, the automated-teller machine, the cellular telephone, and the Internet. Improving infrastructure also seems to work if it is done well, and we've learned, over the past couple of centuries, how to do it well.

All three of these forces have been expanding more or less steadily since World War II; but they have required several generations for their full effects to be realized. And despite the slowdowns on both the productivity and globalization fronts, and despite the challenges inherent in reshaping some of the old infrastructure (particularly in telecommunications), all three are operating in full force today. Their momentum is, in fact, unstoppable. To the extent that they drive economic progress, the boom will inevitably continue.

But all three of these forces are more complex than most of us realize. So it's worth taking a bit of time to examine each of them in more detail.

Productivity Gains: Not Just a Will-o'-the-Wisp

As an economic measure productivity is straightforward: the results per expenditure, usually measured as the dollar value of output per hour worked. But as a conceptual way to understand a system, productivity is both critically important and mysterious. Wealth in society grows when productivity grows; it is the vehicle through which an economy can grow without inflation. But nobody can quite prove what factors propel productivity to grow.

One of the most credible researchers on the subject is economist Robert Gordon of Northwestern University. Gordon was well known at the height of the dot-com boom as a skeptic about the idea of the "new economy"—particularly the premise that computer technology had pushed most businesses into quantum leaps of new and more efficient practice that superceded the understanding of traditional economics. His research suggested that the national productivity numbers were skewed by data from computer businesses themselves, which had *huge* productivity figures. And no wonder, he said; they were taking in unprecedented revenues with relatively small expenses. But it would all be short lived; they were relying on such temporary spurs to business as the initial build-out of the Internet (when every retailer suddenly had to have a Web page) and the Y2K computer-date scare (when many companies tore down and rebuilt their critical computer systems).[27] When you eliminated those factors from the mix, he argued, then the productivity gains of the 1990s were actually much less than those that America had seen in the 1950s.

In his most recent work, however, Gordon has changed his conclusions.[28] Studies of productivity statistics after the end of the dot-com boom have convinced him that productivity went through a genuine sea change in the 1990s. Or, as Gordon himself puts it: "It is too soon to declare that the American productivity growth revival is a will-o'-the-wisp, since the average annual growth rate has been almost as high since the peak of the New Economy boom in mid-2000 as it was from 1995 to 2000."[29]

To appreciate the full story you have to look back to the years between 1950 and 1973, when productivity grew by more than 2.6 percent per year in the United States. This was the heyday of the postwar industrial renaissance, when the interstate highways, commercial air travel, the burgeoning middle-class consumer market, and new mass-production and management techniques transformed large-scale American industry. Then, after 1973, American productivity growth dropped to about 1.4 percent—probably pushed down by the dramatic jump in oil prices, but also by some cultural factors: the rise of mechanistic and bureaucratic management in many large businesses fostered a general cultural distaste for the button-down quality of corporate life. That (plus their own Vietnam-era interests in political causes and then self-development) pushed many bright and innovative young people into other fields of endeavor. People didn't care as much about productivity as they had in the 1950s, and it suffered accordingly.

It wasn't until 1995 that productivity growth moved back to 2.3 percent per year, and it's remained there, or better, ever since. (There are several sources for these figures, and they all vary slightly from each other, but the basic pattern is still the same.) If the difference between 1.4 percent and 2.3 percent seems small, think of it this way: It's the difference between living standards doubling every fifty years, and their doubling every twenty-five.[30] If Gordon had been right, then we would have seen a major slowdown in productivity growth when the dot-com bubble burst. But that didn't happen. Even during the worst months of recession, according to Gordon's own research, our ability to produce more with less labor has continued to improve at a relatively fast rate.

Why did productivity growth come back in 1995? Why did it take so long? And why is it back to stay? There are a number of factors—unprovable as direct influences, but clearly involved in some way.

Technology. The computer and communications revolution increased productivity dramatically by transforming the way business is conducted. The cellular phone, for instance, by making it possible to coordinate activity on the fly, has dramatically improved productivity for anyone who travels. Wasted time and frustration are sharply diminished. Not long ago I was scheduled for a flight within Europe that was canceled because of a thunderstorm. I hired a car instead. I didn't have time to phone before we left, so I spent much of the long ride on my cell phone, rescheduling my appointments. Had cell phones not existed, those meetings would have failed to take place, simply because I couldn't have communicated while en route in the car. And that kind of circumstance is ubiquitous today.

You can find similar benefits from many other technologies that we take for granted now, but that didn't exist in their current form twenty years ago: the credit card (and its quick authorization process), the personal computer, the automated teller machine, the spreadsheet software program, the inexpensive hard disk and rewritable CD, the laser printer, the fax machine, the revolution in instant-printing and document-delivery technologies, electronic mail (and the technologies emerging for ferreting out spam and other unwanted mail), the transatlantic fiber-optic cable, the Internet itself, and the global positioning satellite (GPS) device. In just the last year the sudden boom in "wi-fi" wireless broadband Internet connections, when combined with the World Wide Web and laptop computers, has made the hotels, coffee shops, and airports of the world into effortless hubs of information access; they're as useful as the best research libraries of the previous decade and far more accessible.

None of these technologies led directly to improved productivity when they were first introduced. Historical research has confirmed that it takes at least ten years and sometimes more for a generation of people to come into positions of leadership and organize their companies and societies around these new tools. In the early twentieth century, for instance, the electric motor was a major factor in boosting production at lower cost in manufacturing. Motors were adopted fairly soon in factories after their invention, but it took decades before businesses took full advantage of them by placing smaller motors throughout the shop floor instead of running all the machines from one central motor, connected by belts.[31] Similarly, it took about forty years between the invention of the light bulb and the electrification of the average workplace in the United States—and about eighty years for the average workplace in Europe. To Robert Gordon that head start in productivity is one of the critical differences that led America to outpace Europe after World War II.[32]

You can see the same dynamic in the amount of time it's taken businesses to organize their tracking, planning, and accounting systems to take full advantage of the dynamic communications that spreadsheet programs make possible.

Other factors. New credit and investment vehicles (including the credit line and credit card) made it far easier for entrepreneurs to take bets on productivity-boosting innovations. Financial innovations, such as leveraging and high-yield debt, allowed more efficient formation of investment capital to fuel businesses. Management techniques like "lean manufacturing," "six sigma," "quality improvement," "systems thinking," "self-directed teams," and even "reengineering" (I suppose you could include "scenario planning" here too), made a definite difference in fostering productivity—not least because they helped make the idea of innovative, nonbureaucratic management fashionable again.

Peter Drucker argues that the rise of innovative management is one of the most significant trends of our era.[33] Robert Gordon credits the evolution of retail practice, especially in the "big box" retail stores like Home Depot and Wal-Mart, which continually reduce their own distribution costs and place relentless pressure on manufacturers to keep prices down. And he suggests that labor unions have also helped, by keeping wages high. The higher the cost of labor, the greater the incentive for finding ways to make work more effective and efficient.[34] Pressure placed on business for short-term stock returns has probably played a part, and so has the pressure for some industries—such as the automotive industry—to transform themselves to meet the imperatives of environmental-quality demands by employing emerging technologies like fuel cells.

Gordon is not convinced that productivity rates will stay high. Labor costs, as he points out, tend to follow a cyclical pattern; when times are good, businesses become complacent and begin to hire again, and productivity goes down. Technological innovation can be seen as short lived; the transition from ledger pages to spreadsheets is only made once. But the other factors are *not* cyclical. Continuous improvement, as practiced by leading-edge industries today, is, well, continuous. It never stops improving. The retail stores' pressure on prices similarly never stops; nor does the pressure from investors on share price. Deregulation rarely, if ever, leads to reregulation. And as we'll see in chapters 7 and 8, the productivity boost from technology will (if anything) accelerate as new waves of technology arrive. Molecular engineering of new materials, biotechnology, new energy technologies, small-scale and solid-state manufacturing, precision robotics, and other new-wave innovations will probably have an enormous impact on productivity, and much sooner than many people expect.

Any one of these factors would provide hope for continued productivity growth. With all of them operating simultaneously, productivity growth becomes virtually inevitable. The bottom line: We will probably have one or two more big hiccups, like the current three-to-four-year plateau in economic growth rates, but slower and with greater volatility. Nevertheless, the underlying driver of productivity will not go away.

The genuine surprise, for those watching pessimistically now, is how much economic growth will prove inevitable over the next twenty years. From productivity gains alone we will probably come close to a doubling of the overall standard of living throughout the world within a generation.

Globalization: Maintaining Trust in a Suspicious World

From 1945, when they were created, until around 1990 the institutions of international development and finance—particularly the World Bank, World Trade Organization, and International Monetary Fund—were essentially Cold War organizations. They had been created to bolster capitalist economies so that they would not fall prey, in the future, to either communism or fascism. International trade soon powerfully revived, but the impetus behind the institutions was as much political as economic. It created an array of allies against the Communist bloc, allies whose economies grew so thoroughly interwoven that they could not help but stand together.

For the first twenty years after World War II the scale of international trade was relatively modest. The United States, least damaged by the war, had a share of about 50 percent of the international economy. The rest of the world was catching up, recovering either from the damage from World War II, from ex-colonial status, or both.

Then came the rise of Japan, the growth of the Western European economy, and the ever-increasing importance of international trade. The global economy developed, however, within the context of the Cold War. With the fall of the Soviet Union and the opening up of China to capitalism, the impetus behind globalization shifted. The integration of the worldwide economy was no longer a means to the end of supporting anticommunism. An international economy *was* the end—with democracy and freedom as complements, and the former Iron Curtain countries (plus China) included in it.

One measure of the success of globalization is that the United States now represents only 25 percent of the total global gross product. That's not because the American economy has shrunk; it's because the rest of the world has caught up. In addition, American business is far more tightly integrated with the rest of the world than it once was. America's biotech businesses are owned by Swiss companies; Alaskan oil is controlled by British Petroleum; foreign-owned media companies like Fox (Australian) and Bertelsmann (German) compete with American companies. I don't regard this as a bad thing; its primary effect has been further integration of worldwide enterprise.

It's no coincidence that the 1990s were the heyday of prosperity; globalization breeds success. It is also no coincidence that globalization (and the organizations designed to foster it, particularly the IMF, WTO, and World Bank) became far more controversial than they were during the Cold War. As the Nobel laureate economist, former World Bank insider and current critic Joseph Stiglitz has eloquently noted, the methods of these organizations are heavy handed; their minions are often arrogant; their political accountability is slim; and their mission is all too often still defined by their Cold War heritage. Arguably, where they have succeeded, they have done so because globalization works, not because their specific policies have.

The IMF, World Bank, and WTO are transitional organizations, that have evolved into their current form and role. There is no plausible future, twenty years from now, in which they will act the way they do today. Many of the people who work within them are aware of this. But they (or other organizations supporting market reforms and free trade, the two primary tenets of globalization) will continue to exist. It is not just multinational corporate interests and financial markets that support globalization, but the accumulated experience of the past thirty years. Despite the history of excesses and mistakes, globalization (as Stiglitz puts it) "has helped hundreds of millions of people attain higher standards of living, beyond what they, or most economists, thought imaginable but a short while ago."[35]

A year ago I was not sure that globalization would survive the array of forces against it. These forces included the protesters on the streets of Seattle and elsewhere; the government leaders in Venezuela and Brazil (for example) who had won office on platforms pledging defiance of the instructions of the IMF and other global creditors; the growing distaste for the United States as the "rogue superpower" whose interests are most served by globalization; and the isolation of Russia and China, which still had their histories of mistrust with other nations, particularly the United States. Because of their size and histories the opposition of Russia and China represented one of the biggest obstacles to globalization.

But not the biggest obstacle. That honor was reserved for terrorism—and especially the threat of future terrorism. The attack on the World Trade Center was also a direct attack on world trade in general. It's difficult to develop a financial or logistic infrastructure across international boundaries if you are afraid to fly overseas and allow foreign nationals inside your borders.

These counterforces to globalization are still active. To the extent that they succeed, the Long Boom is slowed down. At the same time, in reaction, the counterforces have provoked counter-counterforces that support globalization. For example, Al Qaeda's terrorist attacks have accelerated the integration of China and Russia into the security framework of the rest of their world, and particularly the United States. The U.S. government has lowered its rhetorical hostility to both nations dramatically since 9/11, and that has been a very positive step. Behind the scenes the geopolitical thrust in favor of globalization has increased. The politics may be divisive; but the ongoing economic trends all favor increased connection. Investment flows, trade flows, and tourism flows among the U.S., China, Russia, Europe, India, and Southeast Asia continue to expand and develop. The political reactions to the U.S.-led invasion of Iraq could slow or even reverse these positive trends. The breakdowns are occurring with nations that (as we'll see in the next chapter) are chronically "disorderly": some of Latin America, Africa, Indonesia, and especially the Middle East.

Ultimately, the quality of globalization depends upon the levels of trust and mistrust—among governments, international businesses, major investors and bankers, and consumers. The more trust, the more connection, and the more connection, the more boost it gives the Long Boom. That's why the U.S. disinterest in international structures, like global criminal courts or environmental treaties, may be dangerous; it diminishes the amount of mutual trust.

At the same time it's very difficult to destroy the amount of trust that has been built over the past fifty years, no matter how mutually suspicious the leaders of nations become. One signal that trust is alive is the ongoing vitality of the North American Free Trade Agreement. Despite a lack of public attention being paid to it within the United States, despite the dangers of narcotics trafficking in Mexico, and despite the Latin American financial crisis, NAFTA is basically regarded as a success. It has woven together the Mexican and U.S. economies in ways that would not permit their being torn apart now. In 1994, when Mexico had its financial crisis, the United States had to come to its rescue; otherwise, the economic effect on the U.S. would have been devastating. Today, there are talks going on to expand NAFTA to parts of Central America and to Chile.

If the nations of the world come to feel that the U.S. can be trusted, then globalization will pick up steam even more rapidly. If the U.S. isolates itself, this will dampen the pace of globalization, but will not halt it. Either way, the Long Boom will reappear. The United States will not become poorer per capita—it will in fact be enriched—but neither will it hold on to its position as 25 percent of the global economy. For one thing, China and India, which between them comprise more than a third of the planet's population, are rapidly developing middle classes of their own. By 2020 China will be a dominant economic power and India won't be far behind.

There will still be half a billion extremely poor people in India, but there will be another half-billion who are living very comfortably. China may have a middle class of a quarter-billion people or so. Those populations, put together, are one and a half times the size of Europe. As they consume goods and services from the rest of the planet, they're going to change the global economy forever.

Infrastructure: Transforming Old Capacities

Improvements in infrastructure "work" to build wealth because they promote productivity. (Think of how much more productive we are because the telephone, electric power grid, and highway system exist.) They also promote globalization. (We take global infrastructure standards for granted, but they make an enormous difference—as anyone knows who has tried to plug in an appliance in another country's electric outlet without the proper adapter. One of the biggest boosts to the productivity of global trade was the containerization of shipping made possible by global standards.)

But infrastructure improvements have other benefits as well. They create a platform for stable, reliable connections among people. Those, in turn, make trade easier. They give people the wherewithal to beat back the harsh vagaries of fate. (A viable insurance industry is a form of infrastructure; without it few businesses could survive the risks of investment in potentially dangerous new technologies or new foreign markets.) Improvements in infrastructure provide a level of comfort that allows people to balance their work with a fulfilling family life, which is the reason many go to work in the first place. (A good day-care system is also a form of infrastructure.) And they provide the kinds of funding and contact necessary for innovation and technological research. (A good university system is a kind of infrastructure too.)

If productivity and globalization are pushing the Long Boom forward, then infrastructure is acting as a governor. The speed and quality of infrastructure development is the primary limiting force on how quickly we return to prosperity. For examples, consider the following forms of infrastructure, and the ways in which they are evolving today to promote (or restrict) economic growth:

Electric power: In 2001 the state of California couldn't buy enough electricity when it was needed. This wasn't a technological problem; it was an infrastructure crisis. The grid of connections among electricity supply and demand sources wasn't robust enough to meet the needs. The same factor limits developing nations; putting electricity grids in place is often the first step toward prosperity, because people need light and motor power to build small businesses or to hold jobs. The grid problem isn't solved, but it is at least recognized, which suggests that this will not be as much of a limiting factor in the future.

(For more about the evolution of electric power infrastructure, see Chapter 7, on the future of energy and the environment.)

Air travel: In some ways this is moving forward, and in some ways back. Momentum is building for an "air taxi" system to replace our current inefficient hub-and-spoke system. To get from, say, Rochester to Cincinnati, people will wait until there is a critical number of passengers, and then (in effect) charter a small plane together to get there. The first aircraft suitable for this kind of service, a four-passenger lightweight jet plane called the Eclipse 500, claims a travel cost of $.56 per mile: low enough so that the trip might cost several hundred dollars for three people, which is competitive with a full-fare airplane ticket, instead of two thousand dollars per passenger, as you might pay to hire a Learjet today. The parent company, Eclipse, was financed by venture capitalists from the personal computer industry; from the beginning it has espoused the idea that airplanes with enough computer capability on board could chart their own flight paths from airport to airport, and not rely on flight paths chosen by a central air traffic control system.

But this "free flight" air-traffic-control system, as it is called, will require significant changes in the existing air-traffic-control infrastructure. And this must happen at the same time that control is getting tighter for security reasons. The need for increased airline security should be a spur to redesigning the entire air-traffic-control system, just as the concerns about the Y2K computer-date "crisis" became a spur to redesigning the large-scale computer-systems infrastructure of most companies. So far that redesign hasn't happened in aviation. Instead, we have palliative security measures with limited effect. Sooner or later the pressures for a new air-traffic-control system will be too great to resist. At that point the Long Boom will get a significant boost forward.

Surface travel: Traffic gridlock, prosaic as it may seem, is perhaps the most pernicious problem holding back economic growth throughout the world. Metropolitan areas are the crucible of economic activity in the world today, in both industrialized and developing nations, but transportation snarls in all too many of them are intolerable. People commute two to three hours each way to travel only twenty-five or thirty miles—a serious drain on both productivity and community life. Cities like Bangkok and Cairo are already frozen in gridlock; London, Rome, New York, and San Francisco are not that far behind them. In most cities the problem has built up so slowly and persistently that people don't realize there is a problem until the breakdown of transit services precipitates a crisis (as is happening in London today, where a new toll is being imposed on every car that enters the city during business hours).

Most of us realize that the current local transportation systems aren't working. But there is no easy solution. Mass transit, for instance, is often held out as the only feasible solution; and light rail or new subways may indeed prove to be the only effective way to transport people in dense, traffic-clogged cities. But these are expensive, controversial, and slow projects; it's taken thirty years to extend Bay Area Rapid Transit (BART) from San Francisco to the airport and San Jose. Moreover, mass transit contradicts the prevailing worldwide trend: toward more personal transportation, in which people maintain control over their starting point, destination, route, and time of travel. Very few people, and no societies, are willing to step back from the idea that an individual should be able to get on a powered vehicle and go somewhere at will. Thus, rapid transit systems only work when they integrate well with a flexible and mobile society—when they run frequently, run to outlying stations with plentiful and inexpensive parking, and connect easily to airports and each other.

The alternative is more roads, which is also not an easy solution. As highway engineers (and many drivers) know well, traffic demand often rises immediately to fill available capacity. A proposed auto bridge between San Francisco and Oakland, dubbed the "Southern Crossing," has never been built because of environmental opposition. If it is ever built, it will be crowded on the first day it opens. But that doesn't mean it shouldn't be built. It means it shouldn't be built in isolation from rapid transit.

All of this requires long-term, accountable financing, which is perhaps the most critical piece of infrastructure involved. Boston's "Big Dig," a major urban transportation project with horrendous cost overruns, shows how vitally important it is to have effective, systematic methods for planning, tracking, and managing these projects. (The Big Dig had none of the above.) Unfortunately, unless there are better-designed organizations set up to manage transportation projects in the future, the Big Dig is not just a present-day boondoggle, but a harbinger of boondoggles to come.

The technological solutions that many had hoped for are also, so far, disappointing. Computer-controlled cars or automated roads are unlikely to take hold; they require too many divergent industries to work together. For *Minority Report* we proposed the way out of the problem would be magnetic levitation vehicles—the vertical highways on which John Anderton (the character played by Tom Cruise) travels in the film. That was the most speculative element of the film. It is going to be difficult to make that technology work; especially going vertically. Second, installing a big new transportation infrastructure in an established city like Washington, D.C., will be extremely difficult. In the film Anderton also travels by Washington

Metro, which deliberately looks the same as today's Metro. That, not futuristic mag-lev, is the likely look of transit in our nation's capital fifty years hence.

Nonetheless, some solution to the gridlock problem is inevitable. We don't know when it will occur, or where it will happen first, but we know that places with well-designed transit and traffic infrastructure will thrive, if only because the best and brightest will want to move there. Those with good transit and good research universities will become the focal centers for the next economic boom.

In that sense the Big Dig, for all its flaws, sets a valuable precedent for visionary planning. At GBN we are involved in some of the planning for some similar systems, and we know they start with questions, not answers. What do the people of the area want? How do they want to travel? How do we make sure the systems work? How do we ensure they are safe and environmentally sound? The answers will vary from locale to locale, but the imperative will not go away.

Financial instruments and corporate governance: The infrastructure of finance has evolved steadily over the past twenty-five years and continues to evolve, opening up new opportunities for wealth creation every year. As Joseph Nocera points out in *A Piece of the Action*, his definitive history of the financial revolution of the 1980s and 1990s, the deregulation of banking and investment in the Carter and Reagan years led to the middle class getting involved in investment and credit in unprecedented ways. Money market funds, hedge funds, 401(k) and other retirement funds, discount brokerages, automated teller machines, on-line banking, and other innovations of the time were often designed and created independently, but they added up to a new kind of infrastructure. Much of the Long Boom was fostered by the increase in investment capital that these changes brought to the market.

There are at least two kinds of new financial infrastructure on the horizon, and both are designed to build trust. The first is land reform in developing nations. Peruvian economist Hernando de Soto is demonstrating that coherent and universal property laws are a prerequisite for eliminating poverty. Because the property laws are arbitrary or nonexistent, he writes, homes and businesses are often held without legal title—so people cannot borrow on them. "The reason capitalism has triumphed in the West and sputtered in the rest of the world is because most of the assets in Western nations have been integrated into one formal representational system."[36]

The second new form of financial infrastructure is evolving out of the corporate governance scandals of 2001–2002, starting (but not ending) with the Sarbanes-Oxley Act. We are likely to see a new and persistent era of regulation, in which corporations will be much more transparent to investors, much more visible to outsiders, and tested much more explicitly.

The World Wide Web and Internet as underlying information environments: These systems will remain the evolving platforms for digital communication worldwide. The Web is particularly remarkable, especially because it was conceived so casually. It is essentially nothing more than a set of software standards for displaying and labeling information. It grew out of a low-level researcher's paper on developing a library system for a Swiss supercomputer center. Nonetheless, it is arguably the most sophisticated information tool in human history, one that deepens and strengthens every day. History will place it alongside the printing press as a transforming infrastructure for human civilization.

I've noticed that my own dependence on the Web as an information source has become ingrained. Even in the midst of far-flung travel, I rely on my ability to sit at a hotel desk, turn on my laptop, and retrieve information on the Web in seconds that would have been impossible to get in the past: either I would never have found it, or it would have been delayed or been of lower quality. Many of the concerns about the Web's quality—that it would be impossible to sift through the flood of triviality, or that it would be hard to judge the quality of information—have turned out to be unimportant. More people have offered high-quality, useful information on the Web—and it is easier to find than anyone had anticipated.

The key to the Web's value, as Art Kleiner has noted,[37] is not the way information is displayed, but the way it is indexed and retrieved. The Web is a directory of human activity—the first such directory in history that is independent of location or family. My son, who is now twelve years old, is a devotee of Lego. He can find people who are skilled Lego builders, and who have posted their plans and instructions on-line. A community of five or ten thousand Lego devotees will accumulate around these sites. Few of them would have had any way to be aware of each other before. Even the on-line "users' groups" and community computer networks of the 1980s were impoverished by comparison.

Tim O'Reilly, founder of one of the most successful computer publishers on both print and Web media, has pointed out that the record industries and studios who are fighting peer-to-peer file transfer software, and labeling it "piracy," are fighting on behalf of only a small proportion of musicians, directors, and writers: "Piracy is a kind of progressive taxation, which may shave a few percentage points off the sales of well-known artists (and I say 'may' because even that point is not proven), in exchange for massive benefits to the far greater number for whom exposure may lead to increased revenues."[38]

There are a hundred thousand rock bands out there with local audiences and no distribution for their recorded music; they'd love to be Napstered. There are a million writers who are concerned not about whether Random House gets the last penny out of Michael Crichton, but about whether there is anyone reading their Weblog. They're delighted to be digitally copyable, because they know that this becomes the vehicle by which they find new subscribers to their ongoing material. In short, to them, file-transfer systems like Napster, Gnutella, and Kazaa aren't piracy, but rather a valued form of infrastructure. O'Reilly argues that sooner or later even record labels will come around to the same perspective:

> I confidently predict that once the music industry provides a service that provides access to all the same songs, freedom from onerous copy-restriction, more accurate metadata, and

other added value, there will be hundreds of millions of paying subscribers. That is, unless they wait too long, in which case, Kazaa itself will start to offer (and charge for) these advantages.[39]

One of the still-unanswered questions is the nature of the business model that will pay for information on the Web. There is a consensus, currently, that people will pay for subscriptions to premium material like Vindigo. But whether the audience for such subscriptions exists is still uncertain. In part that depends on how large the audience becomes; a hundred million people paying $25 per year each in a catchall premium subscription could support a lot of one- or two-person information providers.

The only inevitability is that, somehow, the presence of the Web will lead to redefined business models for publishing and distributing information. That will, in the end, prove an economic boon, not a restriction. It will open up far more opportunities to make money, and it will help fuel the Long Boom. But to make it work there must be one more piece of infrastructure in place. This is the piece that, when it was blocked, triggered the stock market crash of 2000. Until this piece of infrastructure is in place, it's hard to imagine the Long Boom taking off.

The Great Broadband Crisis

It was broadband—or the lack of it—that triggered the bursting of the dot-com bubble. By early 2000 the Federal Communications Commission made it clear that it would not force incumbent local exchange carriers—the four leftover "Baby Bells," namely Verizon, SBC, Qwest, and BellSouth—to open their local lines to providers of "digital subscriber line" channels. This was one of several shortsighted decisions that the FCC has made in recent years. (Their bureaucratic dithering over standards for high-definition TV has helped ensure that HDTV will not be available over cable TV anytime soon.) The FCC is hobbled by a governance structure modeled on the divisions of media in the 1930s—separate bureaus for long-distance telephony, local telephony, international telephony, cable TV, and cellular telephony. There is no bureau at all for the Internet, and satellites fall uneasily under "international telephony." The regulators in each bureau communicate often and easily with the lobbyists related to that bureau, but not at all with the other bureaus. It's a system tailor made to produce regulatory gridlock, which is arguably what many of the lobbyists want.

In the case of broadband, however, the ramifications of gridlock were immense. They gave the "Baby Bells" a virtual (but temporary) monopoly over broadband-to-the-home channels. These companies could go on charging prices above the $25 per month that "feels right" to consumers for broadband service. They could make it difficult for providers to connect service. They could block connections except in fairly proscribed, limited ways. And they could thus short-change the value of the entire Internet.

That's exactly what has happened. The most affordable and widespread method of connection, called "digital subscriber line" service, is a less-than-ideal technology that uses unused sound frequencies on existing telephone wires to send computer signals, instead of putting new fiber-optic cable directly into homes and offices. DSL is still not available in many neighborhoods, and where it *is* available, getting connected is often a logistical ordeal. Some third-party DSL carriers have gone bankrupt; others, like Covad, are engaged in lawsuits with the Baby Bells, claiming they don't get either the access or technical information they need to expand service. The Baby Bells pay lip service to the idea of local broadband, but their executives worry that it will undermine their core business (point-to-point telephone connections) and the value of their existing assets. Therefore, why take the risk? So they don't. The end result: DSL and other forms of local broadband, like cable modem, have penetrated only 10 percent of the homes in the American market. In other wealthy nations the penetration is closer to 30–40 percent. In South Korea it is close to 100 percent, which may become one of South Korea's major sources of competitive advantage in the coming years.

There's a principle of network pricing called Metcalfe's Law, which proposes that the value of a network increases geometrically when the number of users increases arithmetically. That was certainly true of the Internet, up to the point of the FCC policies. On-line advertising, for instance, can't operate effectively at the relatively slow (fifty-six kilobits-per-second) speed of the fastest modems. It requires local broadband: broadband to the home or office. So does any form of video, long files of text, shared audio files (like those in Napster and its progeny), the chains of choice-making that exist in most on-line retail transactions, radio signals, and audio and video telephone signals. Thus, the audiences for all these forms of new media and e-business were suddenly curtailed. Hundreds of business plans had been based on the premise that those audiences would be present. When it became apparent that the infrastructure would not exist for those audiences to be connected, then the equity structure of those business models was untenable. Some of those ventures were clearly based on wishful thinking and deserved to die; but others were worthwhile and didn't get the chance they deserved.

Similarly, the market for the hundreds of miles of long-distance fiber-optic cable that were laid depended on inexpensive local broadband. Those investments were essentially hung out to dry, and the fiber that was laid is being bought up, for pennies on the dollar, by speculative investors. We have already seen some of the fallout from this in the collapse of companies like WorldCom and Global Crossing. But there is probably worse to come. AT & T, once the wealthiest company on the planet, is now in jeopardy; it may not survive the aftermath of the dot-com collapse. If it goes down, it could take with it the 60 percent of the long-distance network that it controls, and many of its suppliers: Alcatel, Northern Telecom, Lucent, and others. If WorldCom cannot reorganize itself, then that could lead to the bankruptcy of UUNET, the WorldCom subsidiary (acquired in the late 1990s) that runs about 40 percent of the backbone of the Internet. The Baby Bells might be the only phone companies left standing, but more likely, the industry being so interdependent, they would probably fail too. The phone system and the Internet would undoubtedly survive, perhaps after being placed in receivership; but meanwhile, technological innovation and new investment in telecommunications would come to a halt. Already, there is almost no investment in either research or implementation of telephone technology. Since telecommunications is the one piece of infrastructure on which the Long Boom most depends, this is already a tragic waste.

And it is all so unnecessary. The United States government has been loath to invest in broadband because of the ideological view that this is a matter for the private sector, and government should not be involved in it. But government has been a critical investor and progenitor of most of the significant pieces of infrastructure in the history of the industrial era, starting with the railroads. Universal telephone service was originally a federal government innovation, justified because it gave people access to emergency services. The interstate highway system was another federal investment; President Dwight Eisenhower had seen firsthand how the autobahn allowed Germans to move their forces quickly in World War II. (The bridge clearance on all American highways is thirteen feet nine inches—the minimum necessary to allow intercontinental ballistic missiles to fit beneath them on truck beds.) And, of course, the Internet itself was originally a computer-linking project sponsored by the Defense Advanced Research Projects Agency.

Imagine if the FCC or Congress were willing to act with similar vision now—to intervene to promote fiber-optic cable to every home and small business in the country. Suppose they came up with some way of subsidizing the initial cost, perhaps with federally guaranteed loans to broadband providers. We would create an Internet system that lived up to the "information superhighway" rhetoric of the early 1990s, with hundreds of millions of people plugged in, using it in ways that could not be imagined in advance.

That piece of infrastructure would become the base on which a truly new economy could be built.

We'll get there without government intervention. The build-out of broadband—not as DSL, but in the far superior form of fiber cable to the home—is inevitable. But it will take one of two things: either time (until, say, 2012 or 2015, during which interval the U.S. will lose even more of its first-mover business advantage) or a crisis. The crisis might take the form of the collapse of AT & T, or the request for a government bailout. If that happens, then it may be time to subsidize cable TV or other types of industries to carry fiber-optic cable to the home, such as electric power companies or water companies, both of which are accustomed to running lines, and which may not fear the Internet as much as the Baby Bells apparently do.

If the U.S. government—or any government—wants to promote economic boom, investing in this particular infrastructure is probably the highest-leverage action they can take.

The Next Business Cycle

As we've seen, though the fundamental driving forces exist for the Long Boom to resume, it may take as long as nine or ten years before we see the stock market soar again. If it follows the pattern of previous markets, it will be level for a while before accelerating gradually, and then hit another period of frenzied speculative growth around 2010. It will take that long for peoples' memories of the 2000 bubble to fade, and for greed to reassert itself again.

During the intervening decade we'll almost certainly experience several major financial crises. The banking system in China is insular enough that a crisis is likely there. There will probably be others, or amplified repeats of the current crises, in Japan and Latin America; we can expect financial crises in India, and in the United States as well. The kinds of business cycles that produce such crises are not bad in themselves. Each one will represent a dramatic increase of productivity, as we push the frontiers of technology and shake out old industries and old ways of doing things.

Because the Long Boom is inevitable, I regard the current debate over the U.S. national debt, and the deficit, to be essentially a nonissue. The debt will increase, but the Long Boom will provide the wherewithal to pay it off. (Of course, borrowing more just to meet everyday expenses, or to pay for a war, would add pressure to the mix.) The social security system will get a break because people will tend to retire much later; the official retirement age may be pushed back to seventy-five by the year 2025.

There is another debate going on now, mostly under the surface, that may affect the Long Boom's longevity. The immediate focal point of the debate is tax cuts, and whether those should favor relatively wealthy people or relatively poor people. But underneath that debate is the concern that American society may be favoring the "haves" at the expense of the "have-nots." Actually, America is very good at taking care of both its "haves" and its "have-nots." The "haves" are the top 20 percent of Americans (in net worth), who live with relative high wealth. The "have-nots" are the bottom 20 percent, who live in poverty. Both groups are on the radar screen of policy-makers; both groups have been targeted as worthy of help, either with tax incentives or subsidies.

It's the 60 percent in the middle who are neglected. The most critical group here is the "second fifth"—the people whose net worth puts them within the 20–40-percent range of the population in income. These are the people who are just one step over the threshold to poverty. They did all right during the dot-com boom; but the issue is what happens now. Factory workers in rural areas, military families, seasonal workers, many self-employed people (such as taxicab drivers and home-care nurses), and many single-parent households fall into this category. Other countries, such as Japan and Germany, do a great job of developing them as productive components of society, but the United States tends to treat them as if they don't exist. For example, they are too poor to afford health insurance but too well off to qualify for Medicare.

The United States suffers as a result. You can see the difference in the quality of service and the general quality of life. These are the people who suffer from environmental health problems, alcoholism, poor schooling, and lack of opportunity. Ironically, the United States—which was founded as a reaction against European class repression—has inadvertently created its own class of people whom it systematically, unconsciously pushes down.

These people matter to the economy far more than is generally recognized. When they are doing well the economy is doing well. During the dot-com boom, for example, these people thrived. Wages went up while the cost of living stayed low. Thus, they made material headway. Some moved from renting apartments to owning small homes. Some moved from odd jobs to regular jobs—for instance, Webvan hired many of them to deliver. They often had opportunities to train for more skilled work; their children had better opportunities as well.

Now, in the last three years, the 20–40-percent group has lost ground. The cost of living has risen, and wages have dropped. Jobs are scarcer. They are disproportionately hurt, and the rest of us are hurt as well, because they are a principal flywheel of the economy. When they are doing well, consumption and investment "trickle up" from them at least as much as it "trickles down" from the wealthy. The well-being of the 20–40-percent group should be an important social and political goal for our country, as well as an economic goal. I suspect, though I can't prove, that the Long Boom will only sustain itself if it raises the horizon for this group. If it doesn't, then the Long Boom, while still inevitable, will fail to fulfil its potential.

What Could Go Wrong?

Is the Long Boom truly inevitable? One possible scenario could cut it short in its tracks—a sustained, difficult, and draining war.

It would have to be an overwhelming war, however—possibly

even disruptive enough to lead to complete economic collapse. Even in a world of conflict and tension there tend to be elements of order that permit global systems to function. Thus, there could be global economic integration that includes the United States, Europe, Russia, China, India, South Africa and Southeast Asia—but not the rest of Africa, or any of Latin America, or most of the Middle East. That would be a shame, but the restricted network would still be enough to propel the Long Boom forward.

In some ways the United States missed a great opportunity to lay the groundwork for a global Long Boom. After the end of the Cold War either the Clinton administration or the George H. W. Bush administration could have tried to foster an orderly set of international relationships. They could have done what Harry Truman did after World War II: build a set of economic and political institutions—the Marshall Plan, the IMF, the World Bank, the United Nations—that (whatever else you may think of them) provided the groundwork for a peace and stability that lasted half a century.

That opportunity ended when the airplanes hit the World Trade Center and Pentagon on September 11. Instead, we have different set of geopolitical tensions to navigate, and an international structure that is arguably unlike anything since the fall of the Roman Empire. That political structure, coexisting with the gradually booming economy, is one of the most surprising inevitabilities around—and the subject of the next chapter.

CHAPTER 5

The Thoroughly New World Order

Which do you hold as a higher moral value—the rule of law or your loyalty to your friends?

For example: Suppose you're riding in a car driven by a friend who is visibly drunk. And the friend hits a pedestrian. A few days later your friend calls you and asks you to testify on his behalf, and to lie: to say he was not inebriated. When you are called, do you tell the truth—because no one, not even your friend, is above the law? Or do you shield your friend, because he really needs you—and what kind of person would you be if even your friends couldn't trust you?

According to "cultural diversity" researchers Alfons Trompenaars and Charles Hampden-Turner, from whom I have borrowed this conundrum,[40] people from different nations often respond to it in different ways. (For more than a decade, Fons and Charles have been surveying businesspeople and other individuals around the world about this and other culturally varying attitudes.) In Russia, for instance, after almost a century of living under the repressive Soviet regime, people tend to be "particularist"—they value particular friendships and relationships far more than any impersonal rule of law that applies to everybody equally. In a totalitarian system relationships are often all you can depend on, while the law can be twisted arbitrarily.

The United States, by contrast, was the first state founded with a modern constitution, and it has been "universalist" ever since. One of the bedrock principles of the country is that everyone in it should be able to depend on the same basic platform of rights and opportunities. As a country of immigrants and people descended from them, Americans tend to trust and respect the rules—which (according to Fons and Charles) is why we have so many lawyers.

People throughout the world have come to expect universalism from the U.S. In the period after World War II in particular, the U.S. was one of the most consistently important proponents of international law, of multinational institutions, and of broad-based treaties like those that curtailed nuclear proliferation and other forms of high-tech attack. These were new bodies of international law that applied to all nations equally. The political right wing within the U.S. was always somewhat uncomfortable with this idea—its members argued that U.S. interests should come first—but it was a quintessentially American concept, this framework of universal law. It fit right in with such other American concepts as democracy, free markets, and the dream of social mobility.

Europe, by contrast, was a far more particularist place. After two World Wars it was a hotbed of regimes locked into petty squabbles with each other, all conscious that no rule of law could override the alliances and loyalties that developed regularly among them.

The leaders of Europe, from Charles de Gaulle to Margaret Thatcher to Silvio Berlusconi, tended to be flamboyant and idiosyncratic individuals who used their individual relationships as vehicles for advancing their national interests, and barely paid even lip service to universal principles. They could afford to act this way because the "big kids" on the playground—the U.S. and USSR—had divided the world between them, and essentially kept the world in a deadlocked military order.

That was then; this is now. The structure of global geopolitics has undergone a cultural reversal so subtle and pervasive that (although it happened at the end of the Cold War) it wasn't really obvious until the George W. Bush administration came to power.

Europe—in confederation with the other prosperous and "orderly nations" of the world—is now the global champion of the universal rule of law. Its power is the "soft power" of moral suasion. America, which was once a universalist nation in a particularist world, has now become a particularist nation in a universalist world. Its power is military and economic—the power of superior force. Its moral authority, which was once its strongest suit, is now (while still vitally important internally) far less of a factor on the international scene. In the words of Robert Kagan, Americans are from Mars and Europeans are from Venus. And then there is a third set of nations, increasingly chaotic and disorderly, in danger of being written off as marginal by the rest of the world. Their power, when they have it, is the power of terrorism. And if that is the only power available to them, they will use it more and more frequently.

This is a New World Order—thoroughly new. It is inevitable, in the sense that it is upon us today and will continue indefinitely for decades into the future. It's emergence has been a is surprise to both politicians and citizens of all three groups of countries. (You can almost imagine everybody waking up one day, scratching their heads, and saying, "How did I get into *this* kind of country?") And the geopolitical future can henceforth only be understood by watching the way in which these three groups of nations—the orderly nations, the disorderly nations, and the United States—interact.

America the Rogue Superpower

"You are either with us or you are with the terrorists," said U.S. President George W. Bush. That line, from a speech made just after the September 11, 2001, attacks on the World Trade Center and the Pentagon, became the core of the Bush Doctrine, America's new geopolitical credo.

The doctrine declares that nations which harbor terrorists are just as guilty as the terrorists themselves; that the United States will henceforth treat such nations as enemies; and that the U.S. will act preemptively against its enemies, instead of waiting to be attacked.

"The United States will continue to make clear," Bush stated in December 2002, "that it reserves the right to respond with overwhelming force—including resort to all of our options—to the use of weapons of mass destruction against the United States, our forces abroad, and friends and allies."[41] In other words, it would not be for some international body of lawmakers, like the United Nations, to decide which countries are terrorist havens and which are not. Even where nuclear weapons ("all of our options") are concerned, the international rule of law is less important than the independent judgment of American leaders. The United States will henceforth decide whether or not a country like Iraq or North Korea represents a threat, based on its own particular relationships and interests, and it will act on that decision—unilaterally if need be.

It is exactly this kind of rhetoric that angers and frightens the people who protest the U.S. invasion of Iraq. Whether they live inside or outside the U.S. itself, they see the Bush administration as having betrayed the deeper, universalist ideals of our country: for example, the ideal that the United States does not enter any other nation unless invited. America is not protecting others from danger; it *is* the danger.

The English commentator Will Hutton was typical of many when he wrote that in order to have confidence in America's position, "we need sophistication, wisdom, the widest coalition possible, legitimacy—and, of course, a willingness to use force if every other avenue has been closed. Instead, we hear the language of preemptive war (which was outlawed by the Versailles Treaty of 1919)—and this from the greatest and most admired democratic republic in the world, a country that has always prided itself on its respect for law, at home and abroad. Europeans expect much, much more from America."[42] War against Iraq, from this point of view, may well be justified, but only if the United States is transparent and "universalist" in its goals: if everyone else feels like America is accountable to the universal rules of engagement.

That is why the international debate over Iraq has turned from a debate over Saddam Hussein's legitimacy to a debate over the legitimacy of American power. Why is Iraq "evil" enough to invade and not Korea, Saudi Arabia, or (depending on your point of view) Israel or Palestine? If the U.S. doesn't have a satisfactory answer, then the rest of the world will feel it is out of control, and that's a much bigger concern to most of them than a vicious but containable Iraq.

And they're right in at least one respect: America is out of control. That doesn't mean it is untrustworthy; but it means that non-Americans will henceforth perceive it as such. The battle over Iraq, important as it is, is outside the scope of this book—it is moving too rapidly, and its outcome is anything but inevitable. But however it plays out, it has already demonstrated the permanent shift in the geopolitical role of the United States. The terrorist attack against the U.S. on September 11, 2001, may have surprised and shocked most American citizens, but I suspect that, when historians look back at this era, this broader geopolitical change will be recognized as even more surprising and shocking. Americans are not accustomed to being alone in the world. We are accustomed to being welcomed, even to being loved, especially by post–World War II Europeans. In our new role love will be hard to come by and respect and maturity will mean everything. We are going to have to get used to being citizens of the first global rogue superpower.

The term *rogue superpower* first appeared in the press during the Carter administration. It referred to the Soviet Union,[43] obviously a rogue superpower under Andropov if ever there was one. Then, in the mid-1990s, the phrase was resurrected as a reference to China.[44] It was only around 1999 that it began to be applied to America, and then only as a hypothetical. Conservative pundits in the U.S. used the term to argue against intervening in places like Rwanda and Kosovo.[45] America, they said, would have to rein in its "expansionism" and "nation-building," or that would lead us to become a rogue superpower. But now the phrase is used mostly on the left, not on the right, and it no longer describes what America *could* be. Look it up on Nexis or the Web, and you will see it used again and again, these days, to describe what America is—and has been since the end of the Cold War.

A rogue superpower is a geopolitical entity with so much power and wealth that it dwarfs every other nation and group in its sphere of activity, and is subject to little or no constraint on its actions. A rogue superpower does not have to be an empire; indeed, it is probably easier to be one without the costs and responsibilities of colonies and subservient states. A rogue superpower can also be basically democratic, open, and well intentioned (indeed, American power stems from the natural advantages that come from 230 years of relatively open, relatively well-intentioned democracy). But from the perspective of people outside the country, even if they agree that America is well intentioned, those intentions are beside the point. Other civilized nations are tremendously constrained today by the international rules and conventions that they have chosen to observe and enter into together. But the United States doesn't play by anybody else's rules. And because we're the biggest, strongest nation on the planet, we may be able to get away with it.

The Bush Doctrine could not have been proclaimed during the Cold War. It would have provoked too many fears, especially the fear of turning the Cold War into a hot war. And the U.S. was not a rogue superpower then; it was the first among a set of "free world" allies banded together to contain and defeat communism. Today, the U.S. wields such political, economic, and military power that friends and enemies are almost irrelevant. No nation in history has ever wielded this kind of unchecked power before. Even China's Han Dynasty and the Roman Empire were partial powers by comparison, with great parts of the world remaining outside their influence. The United States is now to the geopolitical sphere what Microsoft is to the computer industry: hugely successful, triumphant, proud to let everybody know that we're the number-one top dog, and prepared to play rough when we don't like what others do. Nobody much loves Microsoft, and nobody will much love the U.S. over the next few decades.

The Bush Doctrine is a symptom of this shift, and not just a reflection of the particular predilections and assumptions of the people in the current White House. Al Gore has been one of the few politicians to remark publicly about the potential dangers of the Bush Doctrine. But had Gore been elected president in 2000, he would have faced the same pressures: the presence of terrorism, the demand to never again have terrorists or other nations successfully kill civilians on American soil, and the dangers of giving in to international sovereignty. His Gore Doctrine, had he come up with one, might not have resembled the Bush Doctrine in its details, but it probably would have contained many of the same key points—including the singling out of Iraq for censure and invasion. (I've talked to some prominent former members of the Clinton administration since 9/11. They also believe that a Gore administration would have been forced to do something about Saddam Hussein.)

Certainly Gore—or any U.S. president—would have had to adopt the concept that the U.S. is unique; that unlike other nations, it cannot be bound by the rules of international law. A rogue superpower is the most obvious target of that law, just as Microsoft, by virtue of its size and position, is the most obvious computer-industry target of American antitrust law. Thus, even though the U.S. initiated many of these treaties and agencies, even though the Clinton administration paid lip service to them, America has been moving out of them since the end of the Cold War, and especially since the 2000 election, following which the U.S. withdrew from international nuclear disarmament treaties that had lasted for decades, refused to support the International Criminal Court—a permanent court to deal with human rights violations and war crimes—and snubbed the Kyoto treaty on climate change.

This kind of isolation comes with a price. Great opportunities will be lost. The best thing that could happen for the planetary environment would be a planetwide "Eco–International Monetary Fund," investing in environmental initiatives and resolving environmental conflicts. Not a chance. The U.S. has no interest. Nor will we see American participation in treaties over land mines or the Law of the Sea. Moreover, the repudiation of the U.S. by its former allies will continue and accelerate. Long-standing ties with us will be cut. For example, over the next five years or so the U.S. will retreat from nearly all its foreign bases. Okinawa, Guam, and the U.S. bases in Germany and the Middle East will all revert to their host countries (though it is possible we will have bases in Afghanistan and Iraq). Essentially, other countries won't want American troops within their borders. And the U.S. will happily go along—in part because (as we'll see shortly) technology has made those bases somewhat redundant, and in part because American leaders will be glad to get free of the financial burden of the requisite military aid.

At the same time there is another inevitable force to reckon with: the drive within the United States to retain ties with people of other nations. The American people, particularly those who use the Internet regularly, do not want to be citizens of an international pariah. We want to retain the kinds of warmth and mutual respect and influence that we enjoyed in the late 1980s and early 1990s. We want to travel and trade, and to be seen as benign participants in the world scene. And many of us still support and believe in international institutions like the United Nations, if only because there is still a strong universalist streak in American culture. After all, we *still* have a lot of lawyers among us.

This tension within the United States—you could call it the tension between "unilateralists" and "universalists"—goes back to the country's founding. It, too, is doomed to continue indefinitely, and it represents perhaps the most significant uncertainty in America's future. The United States is now the biggest kid on the international playground. Everybody else is at least a head smaller. The U.S. can either be a bully, pushing everyone else around to get its way, or it can be the kind of big kid who watches out for the smaller kids, lifting them up onto the swings and helping them get out of trouble. The U.S. is being pulled in both directions from within. Which side of its personality will win out? It can't avoid being a rogue superpower—there is no nation that can "balance" ours—but the kind of rogue we will be is still not clear.

Ironically, the only enemy that can threaten the U.S. under those circumstances is a rogue enemy like Al Qaeda, suffering under none of the constraints of nationhood—stateless, landless, and unconcerned about the health, well-being, or lives of its people. Al Qaeda can choose weapons that no nation could choose safely, bioterror or chemical, using our own airplanes, using America's own technologies and infrastructure against it. How does the United States protect itself from that kind of asymmetric warfare? It will, in the end, be judged by the way it rises to that challenge.

The Future of the American Military

Before looking at the other "children" on the "playground," it's worth examining the forces that brought the United States into this unique position. The evolution from a tiny democracy—intent on pursuing, as Thomas Jefferson put it, "peace, commerce, and honest friendship with all nations—entangling alliances with none"—to rogue-superpower status did not occur through any deliberate political or military moves. (There were deliberate moves, but by the time they were made, in the years just after World War II, the groundwork for them had been laid long before.) In some respects all this was a natural consequence of America's unique features: its large geographic size, separation from potential rivals by two oceans, openness to immigration, relatively free economy, vast natural resources, and innovative culture. The U.S. has always invited the most ambitious people from other nations to come here; it has treated them well and given them an opportunity to become rich. They (and their children) have included some of the most innovative and capable military leaders and economic thinkers in history, and—from the marquis de Lafayette to Werner von Braun—their ideas have generally been welcomed.

America's superpower status is also a natural consequence of its system of higher education. Bolstered by land grants in the nineteenth-century, and extending the traditional European university privileges to much broader groups of people, America essentially democratized its investment in college-level technological research and development. From the middle of the nineteenth century onward this put America at the forefront of engineering and invention, including military invention. It took about a century for this consistent innovation to develop into a dominant military presence.

Princeton, Harvard, MIT, Stanford, UC Berkeley, and our many other universities are still probably America's greatest military asset. They are autonomous engines of growth that drive the U.S. economy and her military capabilities, in ways that few people recognize. China is trying to replicate them right now, creating Science City and other university centers at tremendous expense. But it's not enough to put the institutions in place. You need tolerance for free speech and free association, and the intellectual climate—down to bookstores, cafés, and bicycle lanes—that attract people to live there. The U.S. is predetermined to retain its dominance in this area, in part because our great universities are already funded through their own endowments and research rewards.

World War I demonstrated that American technological prowess, particularly with aircraft, could make a difference in an otherwise deadlocked major war. World War II tested that prowess much further. Coming out of the Great Depression, with a moribund industrial base, Americans discovered that they were far more powerful than they had imagined. The U.S. led the fighting on two fronts at once, introducing unprecedented amounts of materiel and machines into the war with unprecedented speed, using systems design to coordinate troop movements differently, and building and deploying a weapon of unprecedented destructive capability: the atom bomb.

In the years after World War II nobody knew how far ahead of the Soviet Union the United States was racing—in terms both of military and technological capability. No one else was even in the race. To be sure, the American military proved vulnerable to the Chinese tidal-swarm tactics in Korea and to the Viet Cong's guerrilla warfare, but the defeat in Vietnam led, paradoxically, to even greater military capability for the U.S.; it opened up all of the armed forces to new management approaches and in-depth forms of organizational learning.

The U.S. was already the world leader in military research and development during the Vietnam era, and the gap between us and all other nations has widened ever since. Today, the United States spends at least as much on its armed forces as the military budgets of the next twenty largest countries combined. This was true even during the Clinton administration, when military budgets were stagnant. The American people, by and large, support this; there are few calls for cutting military spending in the U.S. We now have the most innovative military forces on the planet, with the most advanced weaponry and virtually complete freedom to develop it further. (The army, for instance, maintains what is probably the world's most sophisticated simulation system in its computer-enhanced strategic training grounds; one of them, at Fort Irwin, California, has more than six hundred thousand acres and operates at a cost of $1 million per day.)[46]

The biggest change since Vietnam is the enormous technological capability that has come with computer power: fine-grained sensors and remote control allow for unmanned vehicles on the battlefield. The next fighter plane is likely to be the UCAV (unmanned combat air vehicle). It is like a video-game console that operates a genuine aircraft, with the human pilot potentially hundreds of miles away from the front. There are also unprecedented breakthroughs being made in the technological and biological enhancement of individual soldiers: laser eye surgery for fighter pilots, for example, can give them "eagle eye" vision that is far sharper than 20/20. Drugs can allow for seventy-two hours of sleep deprivation with no degradation of performance. Military research has developed enhancements for night vision, pain relief, healing, and general strength and stamina. No other nation comes close to having these capabilities—although, as we saw in Chapter 2, there will also be enormous pressure to release them into the commercial sphere.

American military technological capability is predetermined not just to advance, but to accelerate. This is particularly true when you add the other factor: the Strategic Defense Initiative. Popularly dubbed "Star Wars" in the 1980s, it has been publicly billed as a missile defense system. Naturally, people object to it on those grounds; it doesn't make sense to spend all that money to shoot down incoming missiles when the greatest threat to the U.S. comes from suicide terrorists armed with a rented truck, hijacked airplane, or container ship.

But defense against long-range ballistic missiles is not the main purpose of Star Wars. There is a whole new sub-rosa doctrine of orbital war, not yet well understood by the public, already beginning to play out. The short-run goal is satellite protection. The U.S. economy and military have both become dependent on satellites; those satellites have to be protected from attack by missiles with warheads or by other means. A single nuclear warhead, exploded in orbit, could create a radioactive environment that would degrade every one of the military communications and spy satellites in the sky.

And the long-run goal? A Pax Americana. Star Wars only makes sense as a design for weapons in space for use against the ground. And that, in fact, is its quiet objective: total American military dominance of the planet, in near perpetuity. For example, one of the weapons concepts is called the kinetic spike: a shaft of steel with a nose cone, suspended from a satellite, able to be dropped back down from orbit onto any location on the planet. It's not very complicated; just a steel bar. But when it hits the ground, it will be traveling at seventeen thousand miles per hour, endowing it with enough kinetic energy to destroy a battleship.

What would stop any other country from pursuing the same type of technology? The answer: The U.S. won't let them. Any space battle-station preparations are visible and can be demolished before they get too far along. The system would be virtually invulnerable; it could be destroyed only by its own inattentiveness to some innocent-looking activity on the ground that ultimately became a ground-to-satellite rocket; a stealth form of satellite terrorism.

It's reasonable to think that a fair number of people in the United States would support this kind of powerful system if they truly understood it. After all, should the U.S. manage it successfully, it represents near-complete protection from terrorism or any other threat. It would put the United States permanently into a global sheriff's role.

However, if you believe that absolute power corrupts absolutely, then the potential for corruption in this system is immense. It's not clear that any nation, no matter how benign its origins, should have that kind of complete dominance of the planet. And the Star Wars system has other negative effects as well. It separates the consequences of war from the act of war. It distracts the aerospace industry from its other missions: exploration and commerce.

Nonetheless it is predetermined to proceed—the necessary research and development is already in the pipeline. The latest plausible date that it would be operational is 2020. We may see elements of it in public view sooner, but by 2015 or so its implications will be visible and evident to everyone. A plausible scenario might be imagined whereby the political will is summoned to stop it, but currently there is no political opposition to it at all. To be sure, the Republicans are the most enthusiastic supporters; it is directly related to a vision held by the Bush-Cheney-Rumsfeld wing of the party, a sense of manifest destiny—that it is America's fate and vision to be strong enough to keep the world safe from itself.

But prominent Democrats have not opposed this vision. Even the Clinton administration supported it, at least cautiously. Most battles over Star Wars have focused on the timing, budget implications, and which congressional districts get the jobs. I believe most members of Congress, of both parties, are aware of the implications, and are also aware that if they opposed it, they would be labeled "appeasers" or "soft on terrorism."

At the same time this plan for space war has been evolving below the surface of events, a more visible change in American military culture and intentions has emerged, symbolized by the end of the Powell Doctrine. Colin Powell, who had been a combat officer in the Vietnam War and an aide to the National Security Council in the Reagan White House before becoming chairman of the Joint Chiefs of Staff, established the precept that the U.S. military would not intervene abroad unless it could present overwhelming force. In the past, for example, though the U.S. armed forces included various covert operations units—the Green Berets, the Navy SEALs—they weren't put into action very frequently. Our top military decision-makers were trained to deploy massive forces, like the Seventh Army and the Sixth Fleet.

In the next twenty years we'll see much more police-action-style work by the U.S. military, deploying special forces in other countries. The CIA operative forces, the Green Berets, and the Seals will all be considered predecessors of the special operative forces that will conduct most of America's military work, with the role of conventional forces diminished. As I write this, the U.S. is sending 1,700 troops into the Philippines to scour its southern islands and help rout out 750 members of Abu Sayyaf, a group linked to Al Qaeda.[47] Such deployments will be increasingly typical. Another model will be Mossad, the Israeli intelligence agency. After the 1972 attack on Israeli athletes at the Olympic Games, Mossad tracked down every single member of Black September (the terrorist group that had conducted the attack) and killed him. This sent a clear message, and that kind of attack has never taken place again.

In short, willingly or not, the United States will be drawn into the role of high-tech global policeman. This is a role that the U.S. is not necessarily well suited for, because *nobody* is well suited for it, and nobody has ever had to do it. Politically, the U.S. won't be able to create a police state within its own borders, but it may establish one throughout the rest of the world. It's already moving in that direction; the avowed policy of the U.S. government is that international surveillance is necessary to prevent terrorism. In fact, it may well be.

Can the United States be mature enough to exercise this level of power and not be corrupted by it? This is perhaps the most serious test of the United States on the horizon—and potentially the most serious test of the world as a whole. The current administration believes that, by being tough in a tough world, it can bring order to an otherwise ungovernable situation. If things proceed in a benign, mature fashion, then the U.S. will become more diplomatic and less belligerent in its rhetoric. The rest of the world will come to accept the fact that there are benefits to having one powerful friend to foot the bill for aircraft carriers and missile defense. And we will settle into a relatively benign Pax Americana.

But there is also a worst-case scenario: During the next twenty years the U.S. ends up an isolated hegemon. The Europeans and Asians enter into coalitions, usually led by the French, with the objective of denying the U.S. unfettered power. This would provoke even more belligerence and isolation on the part of the United States. It could lead, for instance, to the kinds of trade wars that would put an end to the Long Boom *and* a continued terrorist presence around the world.

In the end, for all its military might, the most powerful weapon the United States has to wield is the one that seems most unfamiliar to the Bush administration: the cultivation of trust.

The Orderly Nations

Not long ago I conducted a large, in-depth scenario project with the Defense Advanced Research Projects Agency of the United States—the people in the room understood military geopolitics as well as anyone in the world. I asked them: Isn't there any plausible scenario in which a united Europe could become, in effect, a military rival of the United States? What if there was a new European Union president who became, in effect, a twenty-first-century Napoleon or Hitler?

We spent some time exploring both the existing military capacity in those countries—the base on which they would have to build a war machine—and the political environment. I put it in very specific terms: Could we imagine a scenario in which Europe would end up invading the Muslim world? After all, the Muslims are right on their doorstep, there is a history of centuries of mutual enmity, and the pressures of immigration and terrorism are threatening Europe's very existence. Might that not be enough to provoke a new European militarism?

"No chance," said the experts. "Absolutely zero."

Europe has become a postmilitary power. It cannot develop the kind of global military capability that the United States or even China has. It cannot wage war on someone else, even as a united bloc. It is not even equipped to wage war on its Muslim neighbors, which is why migration and terrorism are so challenging. And it is extremely unlikely to wage war internally. Europe will be preoccupied with the integration of its new members and the deepening of political institutions for decades to come.

This is a remarkable transition, one not fully appreciated outside Europe itself. Less than a century ago European nations were killing each other's peoples in very large numbers. The continent in its history had spawned such aggressors as Sparta, Alexandrian Greece, the Roman Empire, the Vikings, the Germanic and Slavic barbarians, Imperial Spain and Britain, Prussia, Napoleonic France, Czarist Russia, Imperial and Nazi Germany, and the Soviet Union and satellites. Europe's very geography, split by forests, mountains, rivers, and channels, had encouraged rivalry and conquest; Europeans' contempt for each other was ingrained in their habits and even their languages. Now they have managed to develop a political process that is binding them together by a set of common rules, legislating their behavior even down to the kinds of sausages and chocolates they sell. And they are not going back because they, and the rest of the world watching them, now have a visible, visceral demonstration of the quality of life that accrues through peace and prosperity.

Of course, Europe has also always had a highly commercial culture, and it's no coincidence that the European Union organizers, with the goal of political union from the very beginning, started to create it through commerce. They began in 1951 with the European Coal and Steel Community, a joint effort by six nations (Belgium, France, West Germany, Italy, Luxembourg, and the Netherlands) to manage the boom-and-bust supply problems of these classically cyclical industries. The first rules of the new Europe regulated the amount of steel or coal each nation could produce.[48]

It took a lot of conversation to work that out. And the Europeans learned something in the process: The way to avoid war is to talk to each other. You conduct elaborate conferences whose subject is the rules for your mutual behavior. You keep talking and talking until you get there. The British, who had been ardent advocates of a European Union (Winston Churchill had called for a "United States of Europe" in 1949), balked at this. To them it smacked of bureaucracy. Margaret Thatcher particularly hated it. But the French believed that was exactly the point—the nations were talking to each other, not killing each other. And they've spent sixty years now mastering the art of rule-making instead of the art of war.

What we have here is the emergence of a new type of nation: the orderly nation. Orderliness doesn't mean internal order; it means willingness to abide by international order and the extensive processes of collaborative rule-making with other nations; participating in a well-ordered set of "rules of the game," not set by a global central authority but continually revised and reconsidered through international deliberative bodies. Some countries that have never had a rule-of-law tradition, like Russia, Indonesia, and even China, are now becoming nations of law. Countries that never had much sense of environmental preservation, human rights, or cultural heritage are now establishing norms and rules for those.

The economic result is visible to everyone: the Orderly Nations have all established stable, self-sufficient economies. South Africa's "orderliness," for example, has made it the leading economic power of sub-Saharan Africa, despite having had to come out from under the inertia of its international isolation during the 1980s. Not all the orderly nations have entrepreneurial, high-growth economies. Most of the European nations, for example, have low-growth economies. But they don't need high growth to enjoy relatively high income, because their population growth is so slow and their GDP is high. They are comfortable with a slower-growth, less entrepreneurial society, so long as it doesn't curtail their egalitarian ideals or basic quality of life.

The inevitable surprise about Europe, in particular, is its introversion—which will continue for at least the next twenty years. The task of forging a single constitutional federation out of such disparate cultures and countries, including the former Iron Curtain satellites, is taking most of the political skill and concentration on the Continent. They are trying to do what the United States did two hundred years ago: to federalize a continent. But this is a much more difficult task, for it involves a hundred times as many people as the U.S. population during the adoption of the Constitution (300 million versus 3 million), along with a much more complex economy, sovereign governments with a much longer and more divisive common history, and more than fifteen primary languages. (The common language of the European Union, as one current joke has it, is "bad English.") There are also huge disparities between rich and poor; fitting the Netherlands and Romania into one economic system will be a challenge in itself. On top of everything else they will also be dealing with the major migration issues described in Chapter 3—issues that they did not expect, want, or plan for. Europe's major institutions, both political and corporate, can be forgiven for being a bit distracted and introverted at this historical moment.

Why do they keep at it? In part they can see the prosperity and stability it's brought; but also, for many politicians (particularly the French), this is the only way they can imagine preventing a unified Germany, with its large population and powerful economy, from once again dominating the continent. The Germans, for their part, have come to recognize that they're vastly better off as part of an integrated Europe than as a dominant player in a nonintegrated Europe.

One surprise, for many observers, is how well it's all working. The new currency, the euro, has remained relatively stable during a difficult economic period. The constitutional convention process is proceeding on schedule, and it is extremely significant; rather than a mere set of treaties, which can be broken unilaterally, this will be the first transnational legally binding constitution. Some of the results are evident. Europe is the world's most attractive tourist destination—a cultural museum, capitalizing on its long heritage in the arts, architecture, and literature. Though the French may hate the thought, they have become the Disneyland for adults. Europe is also becoming a center of innovation in research areas that have a moral dimension; it will almost certainly compete with Japan for leadership in ecological technology, for example. Already, Daimler and Toyota are the two world leaders in designing and producing hybrid engine "green cars."

The entire enterprise of European Union, which began about fifty years ago, may have as long as fifty more years of transition to go before it is a fully stable, unified political entity. Its originating leaders, including Winston Churchill, Jean Monnet, and Jacques Delors, as well as many people who are not well known outside Europe today, will ultimately be regarded in Europe as "founding fathers," much as Franklin, Washington, Adams, and Jefferson are regarded today in the United States.

The rest of the world has taken note, and the remarkable European transformation is spreading. Russia has bought into it, to the extent that it will probably eventually join the European Union. Even China is buying in, using some of the ideas as a model for organizing its own disparate provinces. India is paying attention. The most prosperous smaller nations of the world—Japan, Singapore, Korea, Taiwan, Canada, South Africa, Australia, New Zealand, Malaysia, Thailand, Chile, Costa Rica, Mexico, the Philippines—have pursued a congruent path. They see that it's not enough to eschew war. To avoid fighting, one must actively seek the alternative: commerce, conversation, and common participation in the rule of international law, through deliberative bodies that often supercede one's own sovereignty.

Probably the most significant new advocates of this approach are China and India, two nations whose foreign policies have been practically defined by local enemies. Suddenly, China's historical enmity against Japan, South Korea, and even Taiwan have been replaced by mutual trade and investment, symbolized most dramatically by China's entry into the World Trade Organization and its agreement to abide by international rules on intellectual property. Taiwanese leaders, who feared Chinese invasion in the mid-1990s, are now trying to establish subsidiary companies on the mainland and to establish more direct flights. India, for its part, because it, too, has become an active player in the global high-tech economy, is beginning to make moves to resolve its knife-edge conflict over Kashmir with Pakistan. It is blocked, in part, by the fact that Pakistan is in a very different situation, having made little economic progress in recent years.

Orderliness within an internationally orderly system seems to breed more orderliness. Countries like China could rapidly move from an intransigent, rough-and-ready political stance to one of open participation. It's as if the rest of the world has told China, "You want to participate in the global economy? You want to join the WTO? Here are the rules. You have to have health and safety standards in your factories. You have to curtail software and video piracy. And so on." And to each step China said, "Okay, check. What next?" Each step into the world community makes China more powerful and prosperous than it was before. This new kind of participation in the community of orderly nations has done for China what three decades of pressure from the United States could not do: it has produced a willingness to temper its own iron rule and move, bit by bit, not just toward economic freedom but toward democracy.

It's fortunate that the rule of law will be a factor in tempering China; it will have a dramatically important role in the next twenty years, as the second most powerful economic and military engine in the world. As we saw in Chapter 3, China is rapidly developing what will be the world's largest middle class which ultimately will comprise something like 400 or 500 million people. It will have developed a fair amount of local democracy but, at a national level, will still be mostly authoritarian and centralized. It will have developed a new political system, quasi democratic, quasi corporate, quasi regional, quasi military—a unique form of governance that Kenichi Ohmae likens to a giant corporation: "Chung-hua, Inc."—not unlike pre-Maoist fusions of the public/private boundaries. If not for the support that it receives from international "orderliness," both China and the rest of the world would find it far more difficult to assimilate all this, as I strongly believe the Chinese leaders are aware.

The rest of East and Southeast Asia will also embrace orderliness. It has no choice. These countries suffered through a terribly unfortunate coincidence: the moment of the 1997 financial crisis was also the moment when China's economy accelerated. Production and profits slowed down so much in Thailand, Malaysia, Singapore, Indonesia, and the Philippines, that foreign investment moved immediately to China, which had no financial crisis. Everyone else slowed down, and they haven't been able to catch up since. As a result they are now on a different financial trajectory; they are no longer Asian tigers. Their growth rates have been cut in half, and they make money where they can most easily—as satellites of Chinese industry, occupying small niches in the Chinese orbit. They will become, in effect, like Holland and Switzerland in the orbit of Germany and France. Their international association—ASEAN, the Association of Southeast Asian Nations, was once extremely powerful. Now it has been marginalized. Some Southeast Asian countries, notably Indonesia and Malaysia, will largely reorient their economies to be suppliers of oil and natural gas to China. Singapore will become an entrepreneurial incubator for China, supplying investment banking, R and D, health care, and financial services to Chinese mass producers. It will become the Switzerland of Asia; already its per capita income is $30,000, which is as high as that of Switzerland and the United States.

India will not enjoy the mass-production capability of China, or the array of satellite nations. But it will be far more integrated with the rest of the world economy. It will have almost as many people in the middle class: perhaps 300 million. Since most of the Indian middle class are fluent English speakers, they are (and will continue to be) very effective participants in the global high-technology community, and they will also continue (and perhaps expand) their traditions of prominence in international politics and literature. The greatest challenge facing India today is the one that Mahatma Gandhi foresaw at the time of independence: the ancient conflicts between Muslims and Hindus that have seethed for centuries will still be going on. Unless India can find a way to deal with these tensions, it will suffer a diminished capacity in the global political arena.

Russia will use its transition to an orderly nation to make enormous progress. Many people don't realize how small the Russian economy has become. Its GNP has fallen to a point lower than that of Illinois. It is poised, in short, for a turnaround, and it has three major assets: a vast wealth of natural resources, a highly educated population (with technological capability in particular), and proximity to Europe. The third factor will allow it to exploit the first two. Over the next two decades, especially as it becomes integrated with the European Union, Russia will become much richer. It may well be the primary source of natural gas, for instance, both for Europe and for the growing middle class in China. However, AIDS has the potential to undermine all the growth possibilities in Russia as it saps the energies of the new generation.

The Return of Public Faith in Government

Why haven't nations like Argentina and the Phillipines shared in this move toward orderly prosperity? Preeminently, both Argentina and the Phillipines have missed out on a significant intellectual trend underlying the resurgence of the orderly nations: the return of public faith in government.

In his book *The Cycles of American History* Arthur Schlesinger, Jr., describes how fashions of opinion have swept across American government in a recurring debate: Which is the most effective way to run an economy—with relatively free markets, or with relatively strong government controls? Daniel Yergin and Joseph Stanislaw trace a similarly recurring intellectual debate on a worldwide level in their book *The Commanding Heights*. The balance clearly swung toward free markets in the 1980s and 1990s, and now it has begun to swing back. People have faith in government again.

Some symptoms of this: Belief in deregulation and privatization has declined. The "Washington consensus" (a set of policies in Latin America aimed at privatizing industry and eliminating government services) has been discredited. Corporations have been shown, once again, to be vulnerable to internal corruption unless they are regulated. Even the terrorist attacks have reinforced this, by making people feel, more fervently than ever, that they need some kind of government presence to protect them. George W. Bush, a Republican president elected with an unabashedly antigovernment message, is the first president to create a new government agency (the Department of Homeland Defense) in thirty years. These swings are never complete, and the pendulum will move back again. But this time, in part because of the orderly nations, we can expect faith in government to be more significant, and perhaps to last longer.

Though few people see it, I am convinced of another inevitable surprise on the horizon: The old argument between "government intervention" and "free markets" will grow increasingly irrelevant in the future. We've learned over the course of the last twenty years that it doesn't matter whether the market or the state is in charge. Both are always in charge, anyway. What matters is the competence brought by the particular people trying to do a task, whether they work for a private corporation, a government agency, or even a nonprofit. If you have competent, civic-minded corporations or if you have competent, efficient-minded governments (like Germany) and they are capable in both instances of doing what they intend, then you're likely to get good, efficient infrastructure and services.

The orderly nations are putting this to the test. Because there are so many of them, and because the international rule of law makes them more transparent than they used to be, people can see effectiveness in action. Because of the freedom of migration, people are literally voting with their feet to move to the more socially and economically effective regions. Migration will be particularly important in Europe and China, where the local governments (in Europe's case, the national governments) have a great deal of autonomy but people can easily move from one polity to another. And the orderly nations have remarkably competent governments—particularly France, Germany, Sweden, Singapore, and the Netherlands. Not all of them are brilliant; Japan, which fifteen years ago arguably had the most competent governments in the world, now has a government that is mediocre at best. (It, too, will catch up as soon as the entrenched interests now in charge are either voted out or get too old to stay.)

The most significant exception is the United States: a hotbed of terrible government. With the exception of the military most agencies in Washington are hopelessly outmoded and in many cases getting worse. The federal government has become so big and convoluted that the best creative minds get stifled no matter what happens—and the antigovernment ideology of the current administration only exacerbates the problem. There seems to be neither interest in nor capability for reform. The recent *Columbia* space shuttle disaster was likely a hint that even the best of the government, like NASA, has lost its edge. The various state governments, which are responsible for most genuine infrastructure, are also largely incompetent, but they are under much more pressure to reform. Some of this pressure comes from the recent collapse of tax revenues, which should have been foreseen. Tax revenues, like commodity prices, rise and fall in a naturally cyclical process. Most state controllers warned of this; most were ignored. We will see if they do better in the next cycle.

The United States will suffer accordingly. Government faces much more difficult challenges, and it will require tremendous capability to meet them. We saw in Chapter 4, for instance, that there will be increased demand for revitalized and new infrastructure. But installing new infrastructure in an old city is a terribly expensive and difficult proposition. Look at the difficulty London has had in modernizing its Underground and train lines. Or the Second Avenue subway under construction in New York, which will cost ten times as much as its predecessor. Finally, perhaps the greatest challenge that governments face will require unprecedented skill and deftness—the ability to deal effectively on the international scene. Military prowess still counts in geopolitics, but it no longer trumps all other forms of power. Most of us will learn, with varying degrees of surprise, how many other ways there are to accomplish national goals in the forthcoming geopolitical environment.

The Rise of Soft Power

One of the scenarios we developed for the Defense Department several years ago was called "Soft Power." It presumed that the orderly nations would discover and wield a very different kind of weapon than military force: the use of institutional diplomacy, networks, financial instruments, nonprofits, and corporations as allies, and international rules to achieve political ends. The term *soft power* was coined by Joseph Nye, dean of Harvard's Kennedy School of Government, who defined it as "a country's ability to persuade other nations to comply with its objectives without the use of coercive force."[49] There are, in fact, a lot of ways to exercise and express power without force. Some, like the leadership games and ploys that Niccolò Machiavelli described, are quite old. Others, like the applications of computer models to financial currency and capital investment flows, are quite new. Most of them have been applied more or less unconsciously before now. But now the "orderly nations" are learning to apply them very consciously, because this is the primary form of power they have left.

The French understand soft power very well. They don't necessarily exercise it well. But I expect a raft of French novels to emerge in the next two decades that will reveal the workings of this form of power, in the same way that Tom Clancy's novels illuminate the underpinnings of American military culture. I also expect the use of soft power to become far more sophisticated in forthcoming years. We will become much more self-conscious about such expressions of soft power as technical standards and the messages in our movies and songs.

We have particularly seen soft power wielded against the United States by the Europeans in recent years. Two examples from early 2001 show how the game will be played: the expulsion of the United States from the United Nations Commission on Human Rights, and the European Commission's block to the proposed General Electric–Honeywell alliance. The Europeans essentially told GE and Honeywell that they couldn't merge because the new company would be too powerful as competition for European companies. These represented highly sophisticated uses of networks, information, and ideas (like the American idea of antitrust regulation, only recently imported to Europe) to achieve the strategic goal of curbing the American political and corporate presence. Nor is there any feasible way for America to respond militarily. The deal, in short, is mightier than the sword.

But soft power is not only used to block; it is used to build, and to create possibilities where none existed before. Russia joining the European Union, when it takes place, will be an immense expression of soft power. Europe will then stretch from Dublin to Vladivostock. No military force could ever have accomplished that. The buildup to the Iraq war has shown precisely how powerful "soft power" can be, and how different players can be good at it. The French showed their ability to block movement and exacerbate tension. The British showed their remarkably good ability to ameliorate tension. The Russians and Chinese have both played very well, staking out positions that provide them with good long-term benefits no matter how the war unfolds. Turkey, by contrast, was drawn into the game by its proximity to Iraq and long-standing alliances with the United States and conflict with the Kurds. In trying to maximize its short-term gains (military aid), it has highlighted its inability to resolve the Kurdish conflict, and that will probably play to its disadvantage in the long run.

For there is often—not always—a moral dimension to soft power. It is conducted by elected democracies, and the values of peace, prosperity, fairness, and quality of life tend to have an innate "face validity" that allows them to trump such values as ethnic conflict. A belligerent tyranny like Serbia under Slobodan Milosevic or Zimbabwe under Robert Mugabe (or, for that matter, Iraq under Saddam Hussein) can ignore soft power; it is much weaker against those nations whose leaders are willing to be international pariahs and to accept the disorderliness and poverty that goes with that. But it is extremely potent in dealing with nations that aspire to orderliness, such as Turkey, India, China, and Russia. Each of these nations must deal with its ethnic minorities in a very different way than it would have followed as recently as five years ago, because each of these nations knows that it will need to stand on the moral high ground of the international arena at some point in the near future.

For nations like China and Russia, having a common enemy (in this case radical Islam and terrorism) represents an historic opportunity to enter the world community, and in particular to realign their interests with those of the United States as well as with the orderly nations. And one has to credit all sides—Russia and China, the community of orderly nations as a whole, and the United States—

for taking advantage of this opportunity. If Boris Yeltsin were still the Russian president, it's doubtful he would have handled it as well as Vladimir Putin did.

The orderly nations will not always agree. Tensions among them—between, for example, China and India—will still exist. But these will no longer be ideological differences. They will be conflicts of interest, the seeking of advantage within a larger international system. The larger international system will possess more moral authority than any of its members. Europe lost its moral authority through colonialism; the U.S., through scandals and arrogance; Russia and China, through communism; India and Pakistan, through war; and the developing world and Middle East, through corruption and mistreatment of their own people. Over the next five to ten years there will be far more pressure on the United Nations to live up to its ideals than there has been in the past. The UN will no longer be propelled by majorities of member nations. It will be propelled by the orderly nations. If the UN does not serve their interests, they will create an alternative to it.

The viability of the United Nations, in the end, will depend on its ability to help the orderly nations manage the rogue superpower: the United States. Doing so will require all three entities—the U.S., the orderly nations, and the UN—to evolve beyond their current levels of maturity. I am personally optimistic that they will, primarily because of the way the U.S.-Iraq war developed. The Bush administration, for all its bluster, has shown that they cannot act completely unilaterally. They need support from allies: Turkey, Saudi Arabia, Britain, Russia, and even France and Germany. Consequently they evolved their position, and continued to seek UN support for their military operations against Iraq. The UN process eventually broke down because France chose to push its own agenda, rather than make the Security Council work. And it will have consequences.

First, it has challenged the legitimacy of the Security Council. But the United Nations is more of a moral force in the world because the United States—the nation regarded as the most impatient with the UN and the most to lose from the UN—at least tried to work through it.

Second, it will legitimate the United States, and mitigate its rogue-superpower status somewhat.

Third, it will have further brought a number of countries into an orderly relationship with the U.S. Russia is moving to increase oil production in direct partnership with the U.S. China is not interfering inappropriately with (for instance) Pakistan, and may turn out to be a quiet but powerful force in neutralizing North Korea. The United States is not behaving like a rogue superpower with these former enemies, and the mutual respect among them is evident.

One can even argue that the United States' hard-line stance on Iraq, while unintentional, was most effective as a kind of "tough-love" theater. It took a powerful nation to stand up and say, "It's a tough and dangerous world out there, and since you won't deal with it, we'll deal with it unilaterally." Had they not used incendiary phrases like *axis of evil*, the UN and orderly nations might never have taken the U.S. seriously. This does not mean that the system is predetermined to evolve toward greater maturity; but the engagement with the UN Security Council does represent a significant step forward from the Bush administration's refusal to participate in other international institutions.

Let us hope this continues. Because the United States and the orderly nations need each other. The United States does need legitimacy, or it will become marginalized—and dangerous to itself and others. And the orderly nations need the United States in order to grow and develop economically. It's not that the U.S. is helping them; in fact, even at its most globally integrated, the U.S. is probably going to become more economically competitive than ever, if only because it will see its relative position in the world economy erode a bit. But the orderly nations can't thrive at all without the U.S. military present, to keep the disorderly world at bay.

CHAPTER 6

A Catalog of Disorder

If you have hopes of living in a future safe from terrorist attack, you should put those hopes aside. There is no plausible future in which terrorism has been permanently neutralized—especially when suicide attackers are involved. Osama bin Laden's September 11, 2001, attacks have changed history irretrievably. It is now evident, for the first time, that a small group of determined individuals, so long as some of them are willing to die in the cause, can do enormous damage to great and sophisticated systems. And in so doing they can elevate themselves to power and influence in the arenas they care about.

The damage that Al Qaeda has done to the industrial world went far beyond the physical loss of the World Trade Center buildings, or the loss of life there, at the Pentagon, and in the fields of Pennsylvania. They inflicted a huge wound on the American and global economy, and much of the bill has not yet come due. All of us who live in America *or* the orderly nations will be paying a "bin Laden surtax" for years to come: the extra costs of military spending, the extra time spent at airports and traffic checkpoints, the costs of maintaining those checkpoints (and other new security measures like inspections of container ships), and the extra disruption and anxiety that increased security measures create for all of us. Nor was the damage limited to 9/11 itself; the anthrax attack that followed the airplane attack cost almost nothing to mount, but the five deaths that it inflicted were enough to produce a nationwide panic and to saddle our government with a $17 billion antibioterror program.

The best answer to terrorism is to ignore it. Otherwise, the terrorists win. Unfortunately, ignoring it isn't always possible. And that is the inevitable surprise facing us from a large, virulent, and increasingly disorderly nonindustrial world. For the Middle East, and the terrorism springing from it, are just a small part of a larger picture of disorder.

As we saw in the last chapter, many nations of the world are entering an orderly phase, in which they join together in developing an international base of laws and rules, and prosper as a result. Tragically, however, many nations will be unable to join this community—either because they are too corrupt, too torn by internal conflict, or too moribund. (Most likely, all three.) Not very long ago Western policy-makers assumed that with the proper application of foreign aid and free-market ideology, these nations could be ultimately brought into the industrialized world. Now we have learned that "development" works about as well as colonization. Development only works when a nation chooses to develop itself, because development requires the kinds of policies (property rights, fair opportunity for entrepreneurs, credit availability, quality education, and so on) that require fine-grained implementation, and these can only be established from within, not imposed from above.

Those nations that cannot make the transition are going to decline even further. They are in for a terrible fate. The orderly part of the world will find it increasingly hard to intervene in them. It will be enough, for many political leaders and their constituents, simply to contain the disorder—to confine it to Latin America, the Middle East, sub-Saharan Africa, and parts of Central Asia. When disorder seeps out, it will be in the form of terrorism, crime, disease, and famine—updated versions of the Four Horsemen of the Apocalypse. And it will also emerge in the form of a newly evangelical Christianity *along* with a continued fundamentalist Muslim movement. Both of these religious movements will have an immense cultural impact on the world.

This chapter represents a catalog of the inevitable effects. It begins with terrorism, and then covers religious war, political corruption and revolution, the Israeli-Palestine standoff, massive crime, criminal statehood, narcotics-based civil war, ethnic conflict, and AIDS. The problems are predetermined, but the responses to them are not. There are places in the world right now where the tensions are building to a near breaking point, and they could snap in one of several directions. We need to pay particular attention, as we go through this catalog, to such hot spots: places like Saudi Arabia (where Islamic rebellion is a real possibility), Indonesia (on the verge of being torn apart by ethnic conflict), Mexico (vulnerable to drug wars), and the Caspian Sea. These are places where, if the "horsemen" break through, the consequences could be enormous.

The "horsemen" are already evident in most of the disorderly world. If the industrialized world writes off Africa, Latin America, and the Middle East, that won't be much of a surprise. Can we turn this set of tragedies around? Finding a way to do so is probably one of the greatest challenges that the human race as a whole has ever faced, and it's not certain at all that we will rise to the occasion.

In the March 2003 issue of *Esquire* Thomas Barnett remapped the world of geopolitics along lines very similar to my own:

> Show me where globalization is thick with network connectivity, financial transactions, liberal media flows, and collective security, and I will show you regions featuring stable governments, rising standards of living, and more deaths by suicide than murder. These parts of the world I call the Functioning Core, or Core. But show me where globalization is thinning or just plain absent, and I will show you regions plagued by politically repressive regimes, widespread poverty and disease, routine mass murder, and—most important—the chronic conflicts that incubate the next generation of global terrorists. These parts of the world I call the Nonintegrating Gap, or Gap.

Terrorism
The Power of Unpredictability

Any clear look at the future of terrorism must start with Al Qaeda, the most sophisticated and capable terrorist group in history. But it must not end there, for Al Qaeda represents only one part of the terrorism story.

Enough is known about Al Qaeda, particularly since the invasion of Afghanistan and the subsequent capture of key operatives, that we can say some things about its future with assurance. First, its immediate activities were broadly disrupted after September 11, 2001; it is splintered and scattered far more than its leaders ever expected. Second, the network was probably more extensive than anyone outside it believed at first, and is becoming more extensive all the time, quietly spawning offspring organizations that have varying priorities. To all these networks the damage done to America—while symbolically crucial—is just a means to their real end: to attract and galvanize enough radical Muslims to overthrow the governments of such nations as Egypt, Saudi Arabia, Pakistan, Indonesia, the Philippines, and Turkey. (As we'll see shortly, Al Qaeda is thus a strategically important part of a much larger group of Islamic fundamentalists.)

Third, Al Qaeda achieved a dramatic victory. They demonstrated their power by hurting the United States. This made them a major force in the Islamic world. Their status and power is only enhanced by the fact that the United States went to war with them. As a result Osama bin Laden is now the most powerful man in the Islamic world. Ironically, his face is on T-shirts; his ideas are prominent. His vision of the future grabs the children. His clandestine tapes are immediately broadcast on Al Jazeera, the independent Arabic news network, and people throughout the Muslim world hang on every word. His power is stateless. He doesn't need the power of the nation-state. He can compel people to act simply through his speech, and his network reaches around the world.

Finally, there is no "solution" to Al Qaeda except the capture and execution of its leaders. Execution of bin Laden and other top leaders is inevitably the United States policy. The Americans do not fear turning bin Laden and his cohorts into charismatic martyrs by killing them. They already *are* charismatic martyrs. And capturing them without killing them presents a raft of difficulties. Where, for example, could bin Laden be held as a prisoner, without constant attempts by Islamic terrorists to break him out? Consider the fate of the Al Qaeda leaders currently held as prisoners of war in Guantanamo, Cuba. If this were a war between two nations, the logical next step would be exchange of prisoners. But these prisoners of war have no state to return to and no rules that can constrain them once they are released. They're not going to go back to Saudi Arabia or Afghanistan and become shopkeepers or shepherds, as ordinary POWs would. They will immediately return to terrorism. Their incarceration represents a sore spot in international democracy, but they cannot be released—not now, maybe not ever. They may outlast Al Qaeda itself as felons, remaining imprisoned long past the time when many people in the U.S. will be able to remember much about Al Qaeda itself. And while we have had success capturing some of the leadership, their ultimate fate is unclear.

Because it is not predetermined that Al Qaeda itself will survive the next few years. That depends on something no one, including Osama bin Laden, can predict for sure: the population size of potential future terrorists. Suppose that Al Qaeda has trained and prepared ten thousand young Muslims around the world. Is that the full extent of Al Qaeda's army, which amounts to a relatively narrow cult that will burn out once many have been hunted down? Or is that the tip of the iceberg? Are there potentially 10 million young men prepared to fight and die for radical Islam? If the former, then this particular round of terrorism won't last long. The attacks of 9/11 might have been the peak of their power. During the next few years the U.S. and its allies will kill or capture Al Qaeda's leaders, including (most likely) bin Laden himself. In the process a few more installations will be blown up, with perhaps one or two in the United States. And then it will all wind down. End of war.

But if it represents a rebellion in which millions of people participate, then nothing on the planet will be the same. We will see a belt of opposition rising against the entrenched, conservative (and American-supported) regimes in all these parts of the world. Saudi Arabia, Egypt, Pakistan, Algeria, Iraq, and Libya will all be taken over by radical Islamicists akin to those who seized Iran during the 1980s. (Iran, ironically, will by then be one of the most liberal and Westernized countries in the region.) The "10 million" won't stop there; they would engage in a war against the West that could last thirty years or more.

We should know by mid-2004 or so. If military operations against Al Qaeda are successful (which, barring an unexpected level of American incompetence, they should be), then activity will begin to die out at first. But if new recruits show up, new cells begin to form, and new covert training centers are discovered in new remote places, then we should begin preparing for the longer, much deadlier Al Qaeda scenario.

In any case, terrorism is predetermined to persist. Islamic terrorism will be with us in various forms, but terrorists will not always be Muslims. There may, for example, be a wave of attacks by virulent ecoterrorists—Theodore Kaczynski (the Unabomber) writ large. They might believe they can make a better, simpler world by bringing down technological civilization, using its own weapons against it. They might blow up the Three Gorges Dam in China, for instance, killing two million people downstream in the ensuing floods. Or new forms of ethnic activists may emerge, freeing themselves from a state they perceive as tyrannical. There is no more certain way than terrorism for a group cut off from political power to get its point heard and expand its influence. In that case, the world will resemble England in the 1970s and 1980s, beset by the Irish Republican Army; or Spain facing the Basques; or Israel facing the Palestinians. The terrorists might be rebels against their own governments; a new population of Timothy McVeighs. They might come from a country ravaged with AIDS, seeking to hold part of the world hostage so they can get better medical care for their people (or revenge for a disease that they perceive as having been deliberately sent to murder them). Or they might represent causes and ideas that have barely been heard of before, but that are compelling enough to provoke people to die, and kill others, in their name. After all, almost no one had heard of Aum Shin Rikyo before they bombed the Tokyo subways with sarin nerve gas.

The power of terrorism stems from its unpredictability. Terrorists succeed when they paralyze the rest of us. Should we avoid nightclubs? Subways? Buses? Toll booths? Schools? Disneyland? Or airplanes? Should we take the train or the plane on our next trip? Or not go? Tourism to Bali has vanished already. Should the government establish spot checks of container ships? Trucks? Airports? Or should firefighters be scouring the hills above Denver and Los Angeles during the hottest and driest months, looking for suspicious characters with matches? Should we move out of cities like New York and Washington? And go where? (Oklahoma City, perhaps?) Should we stock up on duct tape and bottled water? And if we take all these precautions, how can we be sure that terrorists won't find some new approach, some form of chemical poison or biological weaponry—or that they simply won't pick ten houses around our country and blow them up one day? Paradoxically, the less the terrorists *do*, the more power they have, because the anxiety from waiting for an attack is more debilitating to an orderly nation than the action of responding.

Nor can we be sure that our actions have had an effect. Since the September 11 attack billions of dollars have been spent on airplane hijacking defenses. But airplane security for the most part has nothing to do with keeping hijackers off airplanes. All the metal detectors and gate checks exist to make passengers (and the administrators accountable for preventing terrorist attacks) feel as if the airplanes are safe. A determined attacker can still get past all the barriers. Who, for example, checked the food cart that came on the plane, or the worker who brought it there? Probably nobody. There's only one serious difference since September 11: Hijackers know that the passengers on board are now likely to fight back. This is not a trivial change; the "shoe bomber," Richard Reid, was apprehended because flight attendants spotted him and half a dozen passengers subdued him. This deterrence will continue to make a difference.

The anxiety about terrorism is inevitable: there is no plausible future that is entirely safe. To be sure, there will be very few large-scale geopolitical threats. But the orderly nations and the United States will be regularly affected by terrorism—both by incidents of terrorism and the disruption involved in protecting ourselves from them. Sometimes the acts will come from networks like Al Qaeda, and sometimes from "rogue states" that are trying to specialize in one form or another of warfare: bioterror, information attack, or sabotage. And of course there will be superempowered individuals with their own grievances. There will be calls for general action. The U.S. military and Interpol will intervene. The terrorists may or may not identify themselves; they may or may not be indicted as activists. We may never know who they were, as in the anthrax attack. But sooner or later someone else will try again.

There is one more danger from terrorism that must be mentioned; it raises the stakes on clarity and consistency from "orderly" governments. There have often been times in history when governments have declined to pursue terrorists because they agree with their goals. One person's terrorist, after all, is another's freedom fighter. The paramilitarists who invaded Cuba during the Bay of Pigs invasion, with CIA help and President Kennedy's blessing, were difficult to distinguish from terrorists and raised the level of tension that ultimately contributed to the Cuban Missile Crisis. Now the United States government has declared war on terrorism. It, and other orderly governments, no longer have the option of initiating such invasions themselves, not without more significant political consequences. In the end the U.S. may recognize that a war against "terrorism," per se, cannot be won. Terrorism is only a technique. The real war is with a very different enemy, an enemy that so far the U.S. has refrained from naming. But it, not terrorism, is the real driving force behind Al Qaeda, and represents a form of disorder in its own right, even should there be no further terrorist attacks. We can call the enemy radical Islam.

Saudi Arabia, Egypt, Pakistan: The Fate of Radical Islam

Religion is a much more difficult opponent in war than mere ideology. Ideology surrenders. When the Soviet Union collapsed, the communists abandoned their ideology without much mourning. They already understood that it had kept them from their goals—modernizing and developing their society. They adopted capitalism, despite their mistrust of it, because they recognized its power to get them what they wanted. Capitalism won the competition of pragmatism; communism lost.

But religion doesn't work that way. Critics like historian Bernard Lewis (author of *What Went Wrong*) argue that the Muslim nations have failed their people by denying them freedom, suppressing innovation, being intolerant, and squelching economic growth. All of that may be true. But to a religiously motivated group, like radical Islamicists, it is beside the point. Nobody asks, "Who'll bring you more success, Jesus or Muhammad?" Instead, they ask, "Which society is serving God?" And when they go to war, they believe that they must persevere until the end. They will die for Allah because they see no difference between the will of God and the will of the religious state.

That is certainly the case with the most belligerent war-making entity on the planet right now: the Islamic fundamentalist movement, of which Al Qaeda is just a part. This movement has been gathering momentum and adherents since the 1930s, and at an accelerated pace since the 1970s. It is fueled by many factors, including the deep vitality of direct religious experience, and also the deep involvement of radical Islamic leaders and clerics in both educating the young and contributing to social welfare. In many Muslim countries, when people get in trouble, there is nowhere else to turn than to the clerics. This is coupled with an intense mistrust of Western pragmatism: a kind of romantic insistence that success on Islamic terms, even if it means death, is far better than progress on Western terms, which feels like a betrayal of God. Thus, we hear Yasser Arafat's wife proclaim on TV that she is disappointed that she doesn't have a son who could become a suicide bomber, rather than (for example) a son who might become an engineer who could bring electric power to her people. And we see Saudi benefactors sending $25,000 to Palestinian families: not to send their children to American or European universities and escape the cycle of poverty they're in, but to blow themselves up in a square in Jerusalem. Those two gestures, and many more, represent a profound rejection of Western ideas of progress.

Most of all, radical Islam is a reaction to the perverse self-indulgence of the last hundred years of Islamic governmental leadership. The unsavory, authoritarian, and corrupt regimes of the Middle East, suspicious of free speech and rife with secret police and informants, have built their tyrannies with the support and collusion of the West. It was a devil's bargain for American and European governments: easy access to oil and anticommunist support in exchange for protection. Following the collapse of Nasserite pan-Arab regimes in the 1970s, the current Middle East regimes have become skillful, over the years, at a disingenuous balancing act; cadging aid from the U.S. or trading with it on one hand, and condemning it to appeal to the "Arab street" on the other. But the Islamic radicals see through this, and it allows them to hold the rhetorical moral high ground in Arab countries: they are the only people who can publicly point out that the governments are authoritarian parasites. The Islamists give voice, in other words, to the resentment that Muslim populations naturally feel.

That is why they have been compared to Martin Luther, who was a radical fundamentalist of his time in the eyes of the papal authorities. And that is why, when given the choice these days, large Muslim populations are voting for the radicals. Even the moderates and the intellectual elites have become more anti-Western and anti-American, and are now supporting many more radical movements, and more broadly political Islam.

Though they are antidemocratic, and tend to dismantle democracy when they get into power, the Islamic movements are skilled at winning elections. They have won landslides in Turkey, Indonesia, Pakistan, and Algeria, where the people would have freely elected an Islamic radical government, if there hadn't been a subsequent military coup. But in most of the Arab world, where there is no such thing as free elections, they are plotting revolution.

Al Qaeda and other terrorists are just a small part of this force, and not necessarily the cutting edge. For it is a long-term and many-faceted trend. First to feel the pressure was Algeria, back in 1962; the country has been enmeshed in civil war ever since between its government and its Islamicists. The first to fall to Islamic revolution was Iran, in 1979. It's typical that when the shah of Iran was exiled, few Iranians mourned—not even those who disagreed with his successor, the intransigent Ayatollah Khomeini. It took twenty years for the progressive forces in this most Western of Muslim countries to regroup. But at least they *have* regrouped. Today, ironically, Iran is one of the most forward-looking, democratic countries in the region.

At least three countries are likely to be the site of attempted Islamic coups in the next ten years: Pakistan, Egypt, and Saudi Arabia. The probability of *one* of them falling is likely enough as to be virtually inevitable. In all three cases the rulers have tried to appease the Islamic insurgency within their boundaries, using money (Saudi Arabia), war with a third party (Pakistan), and rhetoric (Egypt), as well as a kind of benign neglect (all three), often ceding the cultural and educational institutions of the country to the Islamists in exchange for political stability. But the stability can only last so long.

Consider, for example, the dilemma of the House of Saud: the immense, wealthy dynasty that has ruled the Arabian peninsula through seventy years of turmoil, always looking out primarily for its own interests. In 1974 it was relatively easy for the ruling family to support the rest of the country while indulging its own interests—the country had $170 billion in annual oil revenues, and only six million people. But in 2003 Saudi Arabia earned far less in revenues—only $120 billion. And there are now 22 million Saudi Arabians. The country has a monoculture economy; it has never diversified past oil, whose revenues (despite the current spike) are inevitably going to level off or decline. And as we'll see in Chapter 8, oil may lose its energy hegemony for good in the next twenty years. Paradoxically, the higher the price, the sooner that will happen. Meanwhile, the population is growing rapidly; the average family in Saudi Arabia has five children. (In the United States and Europe it's rare to even *find* a family with five children.) As the revenue per person decreases, the Saud family loses its ability to buy off its population. Meanwhile, dissatisfaction with the regime—its corruption, hoarding of power, and lack of transparency—is growing, as is the anomie of millions of teenage Saudis poised to join a workforce that has no jobs to give. It doesn't help that the Saudi leadership is self-deluding. They see themselves as righteous and pious Muslims, who set an example for the nation; but they are also prone to jet off on weekends of wine, women, and shopping.

In Egypt the government has never recovered from its Camp David peace agreement with Israel. In Pakistan, where Al Qaeda and the Taliban still maintain a presence, General Musharraf's government has not recovered from its support of the American invasion of Afghanistan. These countries lack the money to buy off their people, and their income is too dependent on American and European foreign aid to be stable.

Suppose the Islamic radicals succeed, then, in overthrowing the governments of one of these states? What happens then? If the United States government intervenes militarily in any of these three countries directly after a change in regime, it would find itself involved in an ongoing, expensive stalemate. Never popular in the region, the U.S. would now be in the position of putting itself on the line, year after year, to prop up an unpopular regime. That would be practically an invitation to accelerated terrorism. But failure to intervene also invites terrorism. Like Iran under the ayatollah and Afghanistan under the Taliban, these new states would become staging grounds for intolerance, aggressive jihad, and a rejuvenated Al Qaeda.

Saudi Arabia, as the country with the largest oil reserves in the world and the ancestral home of radical Islam, would become a new kind of Islamic power base. It won't stop selling oil; it is too dependent on the revenues. But it will use its income to oppose the West far more aggressively and openly than in the past. Egypt is a linchpin of Middle East anti-Zionism and home to an Islamist movement that has been fighting the current government for more than fifteen years; we could expect a far more openly aggressive return to the old wars against Israel. And Pakistan has nuclear weapons, a long-standing violent border war with India, *and* the remnants of the Afghan Taliban and Al Qaeda within its borders.

Where that would lead isn't quite clear, but one aspect of it *is* clear: The old American policy of supporting status-quo rulers would have to go by the wayside. The U.S. could no longer fight for stability in the region, because stability, by definition, would be anti-American.

Couldn't the United States forestall some of this Islamic rebellion by promoting democracy—using either a regime change in Iraq or a settlement of the Israel-Palestine dilemma? The Bush administration is clearly interested in both approaches. But the first (Iraqi democracy) requires a successful Iraqi invasion. In that part of the world toughness represents a demonstration of virtue. America was tough enough to overthrow Saddam Hussein, and is trying to set up democratic elections and effectively harness together the warring factions that would try to fill the vacuum into some kind of Iraqi democracy, which would set a powerful example for the rest of the region.

This is all plausible. It may even be far along by the time you read this book. A successful invasion of Iraq could be one of the few conceivable events that would prevent this "inevitable surprise" from unraveling so clearly. (As this book goes to press the invasion of Iraq has begun and a military victory has been achieved.)

It is worth remembering, however, that even if the U.S. accomplishes this remarkable feat, the success will largely (in Arab eyes) represent atonement and compensation for American weakness in the 1991 Gulf War. In 1991 the U.S. (and its allies) threw Saddam Hussein out of Kuwait. But the military leaders who led the assault had expected a long, difficult war. It never occurred to them that the war might turn into a quick rout, and there was no game plan prepared for it. So the U.S. blundered. They promised to support the Iraqi dissidents who wanted to rise up against Saddam Hussein. Then, they defeated the Iraqi army so completely that they had to plow the bodies of slain Iraqi soldiers into the ground with their tanks (the so-called "Highway of Death"). After a hundred hours of being slaughtered, many of the Iraqi soldiers got back into their cars and drove toward Baghdad. If the U.S. forces followed them to Baghdad to destroy the regime, we would have exceeded the coalition's UN mandate. President Bush (the first) chose to halt the ground war at 100 hours.

Hussein, in short, escaped by using his own country as a hostage. Then, as the U.S. withdrew, not thinking clearly about the consequences, they left most of their anti-Saddam supporters back in Iraq, with neither an escape route nor protection. And Saddam slaughtered them. American credibility in the Middle East took a dive at that moment from which it has (understandably) never recovered. Those dissidents who were left alive within Iraq in 1991 are (understandably) waiting until America wins the next invasion before revealing themselves.

A successful Iraqi regime change, difficult though it may be, would be easier than a successful Israeli-Palestinian settlement, the quest for which has stymied American presidents since the time of Richard Nixon. Resolution was closest, perhaps, in the late 1990s, when agreement along the lines of the Oslo accord seemed imminent. The mood for peace was so great that it affected the culture on both sides: Israelis and Palestinians collaborated to tape some seventy episodes of the children's TV show *Sesame Street* with Palestinian animators crossing checkpoints to work with their Israeli counterparts, and Muppet characters from each show's story line visiting their friends on the other side.

Nothing like that is possible now. A peace settlement *is* possible, even likely, but it will take the form of an armed stalemate, inevitably leaving both the Israelis and the Palestinians feeling dissatisfied. It would probably involve the wall around Israel growing more and more solid, an internationally chartered government overseeing Jerusalem, a heavy military presence at the borders, the end of many of the Israeli settlements, and the first, halting moves toward a Palestinian state. It may establish order, but no one will be dancing in the streets and thanking the American liberators.

If you want to see plausible prospects for democracy in the Arab world over the next twenty years, you have to look at the one place where democracy has refused to wither—Iran. Despite the theocratic government there are powerful forces in Iran that continually push the balance toward secular, modernized progress, because the Iranian people want it. That is perhaps the most surprising inevitability about the Arab nations: the desire for modernity and progress cannot be stamped out any more than the desire for Islamic orthodoxy can. Saudi Arabia, for instance, has its share of Burger King and McDonald's restaurants, and they are reasonably popular. Many Saudi nationals are middle-class engineers and managers, who have spent a lifetime working for Saudi Aramco or other oil-related companies. They are not anti-American as individuals; the manager who picked me up at the airport on a recent visit had gone to the University of Arkansas. People I met were candid about their own government's problems, and they were as mistrustful of Saddam Hussein as most anyone in the Bush administration.

In countries where the Islamists win, modernity will become the opposition view. As political expert Michael Valajos points out, the most likely path to democracy in the Middle East moves through Islamic fundamentalism, as it did in Iran. Democratic progressivism is most persuasive when it, too, is seen as a reaction against the oppressive status quo. Eventually, he argues, the people will come around, as long as we stop supporting the repressive regimes in the meantime.

In the meantime, however, the Islamists will hold greater amounts of influence and power. And other inevitable developments are about to exacerbate the alienation they feel from America and everything it stands for.

Philippines, Indonesia, Nigeria, Congo: The Coming Religious Wars

"If we look beyond the liberal West," writes Penn State historian Philip Jenkins, "we see that another Christian revolution, quite different from the one being called for in affluent American suburbs and upscale urban parishes, is already in progress. Worldwide, Christianity is actually moving toward supernaturalism and neo-orthodoxy, and in many ways toward the ancient worldview expressed in the New Testament: a vision of Jesus as the embodiment of divine power, who overcomes the evil forces that inflict calamity and sickness upon the human race. In the global South (the areas that we often think of primarily as the Third World) huge and growing Christian populations—currently 480 million in Latin America, 360 million in Africa, and 313 million in Asia, compared with 260 million in North America—now make up what the Catholic scholar Walbert Buhlmann has called "the Third Church," a form of Christianity as distinct as Protestantism or Orthodoxy, and one that is likely to become dominant in the faith. The revolution taking place in Africa, Asia, and Latin America is far more sweeping in its implications than any current shifts in North American religion, whether Catholic or Protestant."[50]

To Jenkins, in short, the "culture wars" in the United States are negligible. Yes, there appears to be a growing number of evangelical Christians in the United States, including the current chief executive; yet, at the same time, the number of people who disavow any religion at all is growing. And yes, this clash probably foreshadows growing political conflict within the United States.[51, 52, 53] But it will be a pale shadow of the political and cultural presence that evangelical, Pentecostal, charismatic Christianity is already beginning to carve out for itself among the poor, disorderly nations of Africa, Latin America, and Asia.

The attraction of Christianity in this world is precisely its focus on worldly and transcendent hope. All of the crises that face people in developing nations: the economic crises, the breakdown of agricultural societies, the move to crowded new cities, the poor quality of life, the rampant crime and corruption, the cutting off of family ties, the advent of AIDS, and the loosening of morality that goes with urbanization—all suggests that there is a devil in the world. The new Christianity includes a growing practice of faith healing and exorcism, which are highly popular in societies without reliable health care or universal education. People who join the religion have reason to believe that they are going to have a good life, if not in this world, then in the hereafter; it is a source of hope in a dark world; and as with Islam in some countries, the only genuine good works they see are those conducted by evangelical Christian missionaries.

Finally, Pentecostal and evangelical Christianity provide the direct experience of God in a community setting. God speaks through their voice. People feel Him in their soul.

And they feel Him in great numbers. In Africa, according to Jenkins, the population of Christians has grown from 10 million in 1900 (about 9 percent of the continent's population) to 360 million (about 45 percent) today. Latin America has had a similar growth rate in charasmatic Christianity.

"The most successful [of these] churches," writes Jenkins, "preach a deep personal faith, communal orthodoxy, mysticism, and puritanism, all founded on obedience to spiritual authority. . . . Whereas Americans imagine a Church freed from hierarchy, superstition, and dogma, Southerners [meaning people in the disorderly nations] look back to one filled with spiritual power and able to exorcise the demonic forces that cause sickness and poverty."[54] They represent a rousing success for Pentecostal and evangelical missionaries, and the scandals over Catholic priest pederasty, so damaging to the Roman Catholic Church in the United States, have not even made a minor dent in the growth of the charismatic, saint-and-miracle-and-daily-prayer-oriented Christian church of this new wave.

The great danger implicit in this trend, of course, is the potential for a genuinely ugly war between radical Muslims and radical Christians. It won't take place in the Middle East, which is predominantly Muslim, or all in all-Christian Latin America. But the deadly possibility is very real in the Congo, Nigeria, Indonesia, or any African or Asian country with large Muslim *and* evangelical Christian populations. "When U.S. soldiers find themselves in the southern Philippines," says Jenkins, "they're walking along one of the key religious fault lines in the world."[55]

Probably the most dangerous "hot spot" for this kind of religious war is Indonesia, a country already torn by conflict between its Muslim population, its Christian population, and its beleaguered leaders trying to hold the country together. The October 2002 bombing of a nightclub in Bali by an Al Qaeda offshoot group served as a wake-up call to the Indonesian government; they now take the threat of Islamic terrorism seriously. But they show no evidence that they understand how to deal with it. In a country like Indonesia, as with Ireland, it is very hard to draw the line between terrorism and civil war. Thus, there is no guarantee that, twenty years from now, Indonesia will be a single nation; already, East Timor has won independence after two decades of struggle because the central government (like the former Soviet Union) didn't have the resources to keep it.

At the same time, Indonesia is a democracy with a strong (if not numerous) middle class and a growing oil and gas industry; many of its people desperately want to establish a more coherent rule of law, join the orderly nations, and develop their economy. Which is the best route for this—to devolve into separate sovereign nations, some of which (dominated by radical Islam) would end up in jihad against the rest, and others of which would become centers of missionary Christianity? Or to remain one country, and risk perpetual civil war and internal terrorism?

Nobody, inside Indonesia or outside, has a reliable answer to this question. That is why the fate of Indonesia is so important. If, by luck or with help, they find an answer that works, it could be a model for many other nations to follow.

Wars between Christians and Muslims are not inevitable—fortunately. But the pressures that might lead to such wars are predetermined to increase. That is why the use of the word *crusade* by the American president was so worrisome; it could be interpreted by Muslim radicals as evidence that the United States itself is joining in their regional rivalries—on the side of Christian radicals. (Worse still, it could be interpreted by the Christians themselves as support for a war against the Muslims.) If the war does begin, it will start as localized, small-scale fighting among isolated groups in places like Indonesia. Only gradually will the local clashes draw national armies in. But if conflict reaches the level where terrorist acts become a commonplace weapon, it will mean world war, drawing in both the United States and even the orderly nations. Such a war is far more plausible to imagine now than it was before September

2001. Arguably, provoking that kind of war has been one of Al Qaeda's deliberate goals.

That's the worst-case scenario. Even now it seems implausible. But world war seemed implausible in 1913 as well. A few terrorist acts, executed against the right targets, could draw the orderly nations into war despite themselves. Such a war would affect most of the other inevitable surprises in this book—for example, the Long Boom would essentially be suspended for the duration of the war, which could go on for thirty to forty years. There would be many uncertainties, but there is one inevitable aspect to a religious war: it would end in stalemate. Religious warriors cannot be defeated. They can be exhausted or even killed, but in their pursuit of heaven on earth they can make life hell for the rest of us, and their cause lives on beyond their death.

Mexico: Colombianization and the Drug Wars

One of the nations that hangs on the cusp between the orderly and the disorderly world is Mexico. Many of the forces at work in Mexico today are working in its favor: its increasing link to the United States, the transformation toward a market economy, and the growing levels of education. The presidential election of Vicente Fox showed that there could be an orderly transition away from the PRI's one-party rule. Together with the North American Free Trade Agreement, this fundamentally changed Mexico's future. (The war of 1848, in a sense, finally ended with NAFTA.) For a long time Mexican leaders correctly felt that the U.S. treated them as second-class citizens. Now, they are equal partners. When they had their financial crisis at the end of 1994 just as the new president, Ernesto Zedillo, was coming into office, the U.S. had no choice but to rescue them, because the economies were so closely integrated. (In a recent conversation at Davos, between former president Clinton and Zedillo, Clinton observed that the U.S. had actually made a profit when Mexico paid back the loan in eighteen months.)

But narcotics smuggling and production in Mexico—primarily cocaine, heroin, and marijuana—threatens all of the gains it has made, and more. Over a decade ago the French writer Edouard Parker coined the word *Colombianization* to mean the downward spiral that had overtaken the once promising nation of Colombia: a haves-and-have-nots economy with little opportunity, a booming illegal narcotic trade, left-wing guerrillas who use that trade to fund insurgent terrorism, a right-wing paramilitary response from the established upper classes, and ultimately a violent civil war with most people (and the country's economy) caught in the middle. Abundant resources and an educated and productive population were simply overwhelmed by these forces; judges, civic leaders, and peacemakers have been kidnapped and killed.

The Colombian story is tragic in itself—but even more tragic as a template for other Latin American nations. Indeed, since becoming attorney general, John Ashcroft has cracked down harder on supply in the war on drugs, and the aftermath of September 11 has increased the checks on flights into the United States. These have made smuggling from Colombia and Peru more difficult, which has made Mexico a more attractive place for narcotics peddlers. At the same time, as Rand Institute studies on the subject have shown, the law of supply and demand applies to illegal narcotics as well as to anything else. When increased police activity tightens the supply, demand remains constant and the price goes up. Thus, the war on drugs in America has increased the incentives for staying in business. Meanwhile, the increased security checks since September 11 have also indicated that smuggling is rampant from Mexico, and that it is often tied to corruption at high levels of Mexican society.

If the war on drugs escalates much further, these trends could escalate as well. Guerrillas may stake out positions in Mexico, adopting the methods of terrorism to keep their operations intact. The United States could soon find a chaotic neighbor, coping with civil war and rampant crime, on its southern border. Conversely, if it can find a way through this challenge, then Mexico could become a model for the rest of Latin America. The Mexican revolutionaries who captured the world's attention in the early nineties have begun to lose ground, and I think the reason is that people are genuinely moving up into the middle class from poverty. They're getting ahead. They see the possibility that their kids will make more progress than they did. And they understand the value of building for the future. The election of Vicente Fox gave many people hope of real and broad progress. Several years into his term the doubts are beginning to grow and it remains to be seen whether those hopes will be realized. If they let all that slip away because drug lords move in, it will be highly tragic indeed.

The Caspian Sea: Criminal States . . . or Entrants into Order?

Home of the Silk Road and the Great Game, the Caspian Sea region was traditionally the primary trade route for caravans making their way from Europe to China and back. In the 1920s, the nations on this route—Azerbaijan, Turkmenistan, Kazakhstan, Tajikistan, and Kyrgyzstan—became part of the Soviet Union, and suddenly their relevance to the global economy dropped dramatically. Then came the fall of the Soviet Union. Now they have joined the roster of states around the world that are run by crime lords, along with Burma, Zimbabwe, the Sudan, and all the nations of central Africa. These are essentially kleptocratic oligarchies in which those in power systematically loot the wealth of the nation for themselves and their gangs.

Recently, the Caspian Sea region has opened up to Western and private development its oil resources, which rival the Persian Gulf's in quantity and quality, and which have far outstripped the initial expectations for the region. And suddenly these lands have begun to enjoy leverage, once again, in the global economy. In 2000 Saudi Arabia tried to raise crude prices, but Kazakstan, a nonentity in the OPEC oil regime a few years before, blocked the deal by expanding its own production.

As the oil market catches on, a series of challenges has emerged for the nations of the Caspian Sea. How fast can they turn their potential oil and gas businesses into actual production and income? Will businesses be willing to deal with the crooked oligarchs in charge? Will the presence of oil revenues prompt any of these nations to reform enough to join the orderly nations? And if not, will the tension between oil supply and crooked government, or the tension between radical Islam and corporate oil profits, lead to war or terrorism?

Oil has been produced there for more than a century, most recently by the Soviet Union. But it's taken a long time to get the modern Caspian oil industry started, despite the keen interest on all sides, because frontier anarchy and criminality is so hard for oil companies to deal with. There were no clear rules for such basic commercial needs as contracts, equity, ownership. There were four possible approaches for building pipelines—west through Turkey, north through Chechnya and Russia, east to China, and south through Iran to the Gulf—and all four were held up by local oligarchs, pirates, hostile governments, or wars. At least one local baron made the pipeline right-of-way contingent on a huge extortion fee that he would personally collect. And yet, if the oil and gas from the region could be tapped more efficiently, it could be an energy lifeline for both Europe and China, and an important balancing factor to the inevitable disorder of the Persian Gulf.

That's one of the reasons why it matters so much whether the nations of Central Asia move toward orderliness or chaos. They are all starting in roughly the same place: with an immense new source of revenue and a set of corrupt, confused governments, sometimes with records of violence against their own people, generally led by crime lords. In the early 1990s the Soviets pulled out and left no bureaucracy in place. These nations all have a variety of ethnic groups within their borders with various tribal identities and rivalries that, in some cases, go back centuries. Their population is primarily Muslim and almost completely uneducated. They are constantly approached by foreign powers—the U.S., the UK, the Europeans, the Chinese, the Russians—with an interest in helping develop and control access to their oil fields. But these nations can't join Europe—the European Union, for better or worse, is not prepared to accept nations that are primarily Muslim, nor so far afield. Nor are they interested in joining the Middle East; Iran, Saudi Arabia and Kuwait will be their oil rivals, not their confederates.

This is a recipe for conflict. Already, several major projects have ground to a halt because Western companies have found the crime, corruption, and extortion so objectionable, and because it is so difficult to establish enough rule of law to do business. Kazakhstan, Tajikistan, and Uzbekistan could all conceivably blow up into civil war, which could cascade easily to the point where one of the local warlords, seeking to control the flow of money, tries to take over the pipelines. And *that*, in turn, if U.S., Russian, and Chinese interests were opposed, could possibly lead to a new World War.

Sub-Saharan Africa: Back to the Nineteenth Century

Order and the rule of law haven't had a chance in Africa in more than twenty years. Since the end of colonization nearly every state below the Sahara (except South Africa) has been ruled by a series of corrupt kleptocrats. Brutal power *is* the law on this continent. And the resulting basic way of life is dedevelopment. There are fewer roads, schools, hospitals, telephones, and electric power generators each year than there were the year before. Literacy and health-care levels are worse each year. Crime goes up year by year (and it was bad to begin with); living well means living behind walls. Most nations there, including much of South Africa, are not safe for outsiders to visit, or even to drive through; carjackings are common.

The rules of development in Africa have collapsed. So have the rules of international order. So has the ability of anyone in the rest of the world to make a difference. In the next few years you will see the last hope of the continent, the educators and innovators who moved there because they loved the people and wanted to make a difference, quietly begin to slip away. Tourists have already left; tourism in Africa is moribund, with only a few brave stragglers prepared to risk the crime and health risks. Investment has also dried up; there might be money available to invest, but no businesses with the expertise to use the money wisely. Exports are way down. Infrastructure is rotting. Natural resources remain, but few companies are willing to extract them. I did a scenario project for a copper company that had spent $5 million on licenses to develop an immense deposit in the Congo near the border with Rwanda. It was wasted money for them, because they could imagine no scenario in which they might develop that project. Projects like that all over Africa are being abandoned, with the economy deteriorating accordingly. It's even difficult for some nations, like Uganda, to send ambassadors to Europe or the United Nations. They have no difficulty finding candidates; but it's impossible to get them to come back once they leave.

Twenty-first-century Africa, in short, is returning to the technological levels and way of life of the nineteenth century. People are leaving the cities and dropping out of the global economy. They are returning to subsistence agriculture, but now with far greater populations to support. The loss of life and community to ethnic cleansing and other forms of wholesale violence is staggering. Famine and disease—particularly AIDS, as we will see shortly—are playing a major role as well. Hundreds of millions of Africans will die before their time during the next twenty years. The overall population of the continent will probably decline from the combination of disease, warfare, and starvation.

This will ultimately be seen as one of the worst human tragedies in history. The rest of the world will ultimately be seen as having turned its back on Africa—in effect, walling off the continent and its problems. This indifference may well be a form of racism. Or it may simply be indifference: sub-Saharan Africans are remote from every other continent, with nothing much to offer them economically, and little or no ability to touch anyone's lives except their own. Either way the indifference borders on criminal—except, if the outside world were to intervene, where would they start?

A few nations have the potential of climbing out into some form of orderly survival: South Africa (if it can manage the AIDS problem), Ghana, and perhaps Mozambique. Otherwise, one can imagine things getting so bad that there is a move to recolonize the continent and start over, perhaps under the auspices of the United Nations. But it would require at least twenty or thirty years of misery to get to that point.

AIDS in Africa, China, Russia, and India: The Final Barrier to Order

The number itself has a terrible power. Fourteen million orphans. Fourteen million people growing up without parents in Africa, a continent where community and family life are paramount, where there are no resources for taking care of rootless people, where infrastructure (water, sanitation, electricity, education) is minimal to nonexistent, and where mourning will last a lifetime. We have not begun to deal with the consequences of this terrible inevitable surprise: 14 million parentless children growing up in a continent this poor.

But as political economist Nicholas Eberstadt reported in *Foreign Affairs* recently,[56] while Africa has been devastated in medical and human terms by the disease, this also represents just a small harbinger of the devastation to come. AIDS is rapidly becoming a presence in three countries that have enormous influence on the rest of the world: Russia, China, and India. It represents the most significant wild card affecting the future of these countries—especially their ability to become orderly nations.

If human beings were purely rational creatures, AIDS would not be a difficult disease to control. It only spreads through the exchange of blood—either through anal sex, through the sharing of needles, or through blood transfusion. All three forms of exchange are preventable, *if* there is community or political will to do it. And that is the tragedy of AIDS. In all the groups where it takes hold, finding the community and political will is so difficult that denial is practically inevitable. In the United States, for example, AIDS and HIV infection are repeatedly pronounced "cured," because gay men, as a whole, adopt safe-sex practices after they realize the seriousness of the disease. But then a new generation of gay men comes of age. Those who move into a "gay-identified" community (like San Francisco's Castro or New York's West Village) have often chosen a path in which their lives revolve around sexual identity. They are predisposed, heart and soul, to ignore the fact that their life choice has a fatal error in it. Like many young people they cannot believe they, too, would be vulnerable, especially to a disease with a long gestation period. They rationalize the problem; they forget about safe-sex practices every now and then; and the cycle of death starts up again.

Something similar, but far more virulent, has taken place in Africa. Many African cultures accept promiscuous sexual play as a normal part of adolescence, a way of life before marriage. When I was a Peace Corps educator in Ghana in 1968, the other teachers and I were sometimes offered sexual favors by the young women who cleaned our houses. I never accepted, because I didn't think it was appropriate, but neither did I condemn the practice for others. At the time it was seen as a form of routine consensual courtesy. Girls grew up being trained, by their mothers, that nontraditional sexual practices were the most effective (and inexpensive) way to prevent unwanted pregnancies. Condoms in particular were mistrusted, because they had been introduced in the context of population control. They were seen, in some ways correctly, as a Western plot to reduce the number of African people. At the same time, in a culture where relatively few people were literate, the mechanisms of disease were not well understood. Traditional medicine, which often involved faith healing and prayer, was accepted far more than the kinds of medicine that depend on understanding microorganisms and viruses.

Then came AIDS. And the first reaction in Africa was denial as well. They denied it was caused by sexual contact. Or they accused AIDS of being a Western plot to attack African pleasure (at first) or destroy African lives (when the consequences of the disease became clearer). South African president Thabo Mbeki famously announced that AIDS was caused by poverty. To be sure, poverty exacerbates the problem, because it generally leads to lower nutrition levels, and those make the body more susceptible to disease; but focusing on poverty allowed South Africa to ignore the kinds of educational and infrastructure changes that would have helped fight the disease.

There are African men, for instance, who leave home to work in the mines for months at a time, living in dormitories. They have sex with each other during that period, a practice never spoken about overtly. Then they return home to their wives. The wives contract AIDS. To curtail such practices would be extremely difficult; they have gone on for generations. It is perceived to be much easier to revert to faith healing as a way of dealing with it. I have been told by established African intellectuals that "you guys in the West don't understand this disease. It's not sexual at all. We've been dealing with it for centuries."

Contrast this attitude with that of Thailand, where prostitution tourism had long been an established (if somewhat illicit) business, and where 25 percent of Thai men were estimated to be customers as well. When it became clear, in the mid-1990s, that AIDS would be a serious problem in Thailand, the government instituted a series of safe-sex practices: promoting condom use, campaigning against prostitution, and monitoring the needles in drug use. AIDS has not been eradicated in Thailand by any means, but it is being managed; the number of new cases dropped from twenty-three thousand to nine thousand between 2000 and 2001.[57]

Many other parts of the world are at risk from dramatically rising numbers of AIDS incidents. Latin America, for example, had 1.4 million adults and children estimated to be living with AIDS or HIV in 2001. But three countries are particularly important, because of the influence that they have on the rest of the world: China, India, and Russia. All three are at risk from dramatically rising numbers of AIDS cases: 58 million in China, 85 million in India, and 12 million in Russia. In all three cases, as with Africa and Thailand, the spread of the disease is exacerbated by social factors ingrained in the commercial practices and culture. In India, as with Thailand, the exacerbating factor is prostitution—along with the networks of commercial truckers who form a critical part of the transportation infrastructure, and who frequent prostitutes regularly. In China, AIDS is accelerating through the migration factors described in Chapter 3: a large population of unmarriageable males and a massive migration of people into cities. There is one other key factor: a health care system geared toward rapid development, which is therefore careless about blood transfusion. In Russia the flywheel factor is the prison system. "Prison camps," writes Eberstadt, "are virtual incubation dishes for diseases such as drug-resistant tuberculosis and HIV." When released back into society, prisoners carry the diseases with them.

AIDS will be terrible in China and India. Millions of people will die and be orphaned. But the numbers also need to be put in context. Fifty-eight million in China and 85 million in India represent only .05 percent and .07 percent, respectively, out of the 1.3 billion people who live in each of those countries. Russia's 12 million AIDS sufferers constitute almost 10 percent of the country's population. Moreover, the disease is disproportionately centered in the relatively young working population, those whose efforts carry the productive aspects of the economy forward. Moreover, the government has been in denial so far. Unless they tackle the AIDS problem as Thailand did, Russia might not succeed in making a full transition into a modern, orderly, democratic, capitalist, market-oriented, progressive society.

If that happened, Russia would become the largest disorderly state in the world. It would be armed with nuclear weapons. It would have borders with both Europe and China. And it would be, more likely than not, ruled by organized crime.

It would be nice to think AIDS might provide a trigger to push emerging societies past an inflection point on health care: shock them into developing better standards of care, cleanliness, health practice, and sanitation. But tainted blood has been an issue before in many countries, with respect to hepatitis C and other blood-borne diseases, and it hasn't made a difference. There hasn't been money, and as Nicolas Eberstadt points out, tragic as it may be, these societies can't afford the cost of rescuing human lives. The human lives are not worth the $600 it costs to put an AIDS patient on life-maintaining drugs, even generic drugs. It would also be nice to think that AIDS could represent a check on criminal behavior; a way to show the rest of Russia, for instance, that its tolerance of criminal oligarchs bears a terrible price. Here again, the past is not very encouraging.

But it is not predetermined that the future will be limited by the past.

The Choice between Order and Disorder

This is a sobering catalog. Disorder and its terrible effects are inevitable. Some places, like Africa, are almost inevitably doomed to suffer disproportionately. Others have more uncertain future. Most Latin American nations, for example, regularly cycle back and forth between increasing order and increasing disorder. The next twenty years may determine which end up in which camp, once and for all.

It's impossible to predict which countries will navigate the transition, but there are some important leading indicators to watch. For example, how well is the country dealing with AIDS? Thailand's effective approach signals that it is poised to enter the orderly world; it may become one of the best-governed countries in Asia over the next ten to twenty years. South Africa, by contrast, which has gotten so many other things right following the fall of apartheid, demonstrates through its AIDS policy how vulnerable the new "rainbow nation" may still be.

Another good indicator is the behavior of overseas students. Enormous numbers of students go to the United States to study, from all over the world. How many of them return to live in their countries of origin? South Korea was one of the first places to attract its students home, to work in the electronics labs of Samsung and other companies. Now, in China and India, this indicator has begun to shift as well; many more students are returning home. There are jobs and opportunities for them. So long as that remains true, the outlook for China and India is much brighter. Conversely, students from Muslim countries who study science in the U.S. tend to stay there; there aren't many great universities in a country like Egypt, which locks up its intellectuals. That means Egypt is unlikely to develop much of an economic engine of change.

Finally, the behavior of the American government toward various nations is a strong indicator. The United States, because of its wealth and military power, has a disproportionate stake in how many nations become orderly, versus how many become chaotic. If the tensions between the United States and other nations grow, so will the amount of disorder. The battles between the U.S. and other orderly nations will be fought as the Cold War was, in surrogate conflicts in the disorderly world. If the disorderly world is not attended to, if poverty is not addressed, or environmental decay, access to water, health, AIDS, and so on, the failure will become an even greater source of disruption and will dominate the attention of the world.

If, on the other hand, the United States and the other participants in the orderly world work together, to increase integration, to help bring more countries out of the disorderly world, and to feed that process of global acceleration and integration, one can imagine a more benign future.

There is always reason to hope, because there are factors pushing the United States to be more open and resilient as well. One of those factors has always been a strong point in America, since the days of Benjamin Franklin: science and technology.

CHAPTER 7

Breakthroughs in Breaking Through: Science and Technology

There's no mystery about scientific revolutions. It is clear, from observing the revolutions of the past, what conditions are necessary to create them. There are four basic factors that pave the way for major breakthroughs in science and technology. They were all present in the early 1600s, the time of Nicolaus Copernicus, Johannes Kepler, and Galileo Galilei; they were present again at the turn of the twentieth century, the time of Thomas Edison, the Wright brothers, Guglielmo Marconi, Albert Einstein, Max Planck, and Niels Bohr.

And all four factors are again present now.

The first is the emergence of scientific anomalies: new discrepancies and paradoxes in old scientific models, which are raised when new facts come to the surface. When Copernicus, Kepler, and Galileo questioned the prevailing Ptolemaic model of Earth at the center of the universe, or when Einstein and Bohr found inconsistencies in Newtonian physics and replaced them with another model of physical reality, the repercussions included vast new possibilities in science and technology. Without quantum physics, for example, we would not have had solid-state electronics, lasers, or nuclear power.

One anomaly emerged in 2000, when a group of astrophysical researchers discovered that the universe is expanding at an accelerating rate.[58] This contradicted the prevailing theories about the nature of gravity; for if gravity represents an innate attraction of mass to other mass, then it would theoretically act as a brake on the expansion of the universe.

To make sense of this some physicists posited that some cosmic force exists to overcome the influence of gravity. They have called this entity "dark energy." But where would dark energy come from? Why would it exist? How would it operate? There are no satisfying explanations yet. Our current assumptions about dark energy are much like Ptolemaic epicycles: the elaborate, imaginary cosmic wheel-and-gear machinery that medieval astronomers devised when they needed to reconcile their theory of the universe—the Earth at its center, and the planets in perfectly circular orbits—with the actual astronomical motions they observed. When the truth about the Earth spinning around the sun became known, one of the biggest points that persuaded people to accept it was the tremendous relief they felt because it accommodated the previously unaccountable data. They could now discard those horribly intricate and ungainly epicycles.[59] And there's a similar feeling about "dark energy," even among the physicists who proposed it. At the first big meeting of the American Physical Society on dark energy, a physicist from the University of Chicago said, "Well, this clearly shows we live in a preposterous universe."

The universe may seem preposterous, but its workings likely have a coherent explanation that we just haven't found yet. If some energy, unknown to us, is vastly greater than all the gravitational force and all the mass of the universe, then that suggests the need for a shift in the prevailing theory about the nature of energy and matter—which in turn may have immense consequences for our scientific understanding in the future. Interestingly, Einstein's early work foreshadowed this anomaly. His equation of general relativity included a factor he called the "cosmological constant." In his forties, he took it out of the formula. He called it his biggest mistake. It turns out he may have been right the first time; the cosmological constant produces the same behavior that is evident in dark energy.

The "dark energy" anomaly is just one of several that have surfaced recently in physics, chemistry, biology, and "multidimensional" mathematics. At the subatomic level three different kinds of neutrinos have been identified, and two of them have mass. That contradicts all previous atomic theory: every college physics student from a generation ago would know that neutrinos don't have mass. But now we believe some do. We don't know which of these anomalies will be resolved in the near future, but the presence of so many represents the clear signal of an impending breakthrough.

The second basic factor required for a scientific breakthrough is the development of new instruments that detect phenomena never observed before. Indeed, this is where scientific anomalies often come from. The telescope famously empowered Kepler (who relied on telescope data collected by his mentor, Tycho Brahe) and Galileo; the particle accelerator ("atom smasher") provided the empirical data for much of the "new physics" of the mid-twentieth century. This time around the discovery of the universe's accelerating expansion was made with instruments positioned on satellites: orbital telescopes and devices for detecting X rays and gamma rays. When trained on distant supernovae, they recorded the impact of the accelerating universe on the apparent brightness of the exploding star.

Another telescope is planned for 2012 that will remain in orbit at the "L2" point, a position in Earth's orbit beyond the moon. It will be able to detect planets the size of Earth as distant as three hundred light-years away. A subsequent telescope will be able to create images of the surfaces of those planets. Meanwhile, other new instruments continue to push the edge of measurement at the subatomic level. IBM recently announced a new electron microscope that can detect and distinguish electrons. And there are yet other new instruments that deconstruct time. Femtosecond cameras can capture images of processes that last only 10^{-15} seconds. At that speed you can see all the steps of a chemical reaction unfold, watching subatomic particles swirl into place, one by one, to align in the formation of a new substance. It may open up the ability to manipulate chemical reactions, atom by atom, to create substances from scratch.

Other kinds of sensors and instruments are also under development. One set of instruments uses terahertz wavelengths, which are a little bit shorter than ultraviolet light. At low power they can penetrate living tissue without damaging it. This technology may well bring us tremendous information about all living creatures, starting with medical information and knowledge about trees and forests. Finally, as we'll see shortly, immensely powerful new computer technologies are inevitably emerging from quantum physics research and biological research, which promise to enable such large-volume computational feats as gene processing.

The third factor is rapid and effective communication among scientists, especially in comparison to the past. The printing press was still relatively new in the seventeenth century, but it was sufficiently well established to permit the circulation of scientific papers and work. Galileo and Kepler knew of each others' work, as did the other leading scientists of their time (including those who were officials of the Church, which led to Galileo's trial for heresy). In the late nineteenth century, innovations in telephony, telegraphy, and printing were in place; not only did scientists use them to communicate with each other, but their discoveries were keenly followed for the first time by mass audiences.

Today, of course, the Internet has revolutionized scientific communication. Peer review no longer requires delays of six months to two years; it happens almost instantaneously. Scientific journals are morphing into free-form Web-based publications, and the scientific conversations on the Web have evolved into a sort of ongoing, nondisruptive, perpetual colloquium, allowing for connections and collaborations that would never have been possible before.

The fourth factor is a political and economic culture that values science and technological research, and rewards people for it. In eras when individuals can get rich and countries can become powerful through science, research flourishes. Science and technology were sponsored in the seventeenth century by Italian nobility and other patrons; and in the nineteenth century by a broad new wave of technological investors and government sponsors (both in the United States and England) who sensed that there were fortunes to be made.

Today, the United States has probably the most lavish funding base for science and technology in the history of the world. It funds research through venture capital, government funding ($75 billion annually from the federal government), corporate R and D, foundation grants, and university endowments. Billions of dollars flow through these channels for basic science research. Often not targeted to specific goals, and therefore unpredictable in its effect, this is the hardest kind of funding for politicians in most nations to justify. Nonetheless, it is often the source of the most significant breakthroughs. Our current drugs for treating anthrax, for example, come not from an intensive effort to protect Americans from the disease (though the military funded such research for years), but from a researcher's longstanding curiosity about the structure of toxins. The National Institutes of Health funded his research for fifteen years before there was any practical outcome.[60] A related key factor is the American network of research universities, which (as noted in Chapter 5) provides an essential infrastructure for both channeling the money and conducting the research. Other nations like China, noticing this, have begun to develop technological research centers of their own. It will take them a long time to catch up, if they ever can; but in the meantime the competition will spur the U.S. to even more intensive basic R and D investment.

Those four factors, combined together, create a kind of momentum that is increasingly recognized by scientists themselves. The National Academy of Sciences, after a survey of the field of physics, recently concluded that it is ripe for a revolution. That perception, in turn, has spurred even more activity, as scientists vying in their particular specialties vow to be first to publish their results. Which, in turn, spurs more investment. As a community of investors and researchers we are increasingly creating the scientific revolution that we collectively see as inevitable.

By definition it is impossible to provide all the details of a scientific revolution that is not yet born. But some of the arenas in which breakthroughs are likely to take place are already evident, and we can speculate on some of their end results. Moreover, we have a fairly good idea of the pace of change to come. There are three distinct types of inevitabilities, each with its own time frame.

First will come the breakthroughs to emerge from research that is already under way. It is relatively easy to tell which technologies will emerge, and some of their effects, while perhaps surprising, are also predetermined.

Second will come the breakthroughs at the frontiers of today's science. In this book I can point to arenas where breakthroughs are expected, and identify some of the ramifications and issues that are raised.

Third will be the paradigm-changing breakthroughs, after which science and technology are never the same. The form these might take is unknown—we can only speculate. But there is an inevitability here too. In fifty years' time knowledge of physics, biology, chemistry, astronomy, and maybe earth science will be immensely different from knowledge today—far more different from today's knowledge than ours is from that of fifty years ago.

Stage One: Small Systems, Few Secrets

Perhaps the most surprising breakthroughs to come from research already under way have to do with thinking small. When Eric Drexler first popularized the concept of nanotechnology in his 1987 book *Engines of Creation*, it hardly seemed plausible—manufacturing and robotics on a tiny scale, down to molecular-level manipulation. Now it is becoming mainstream. Chemical, biological, and material systems are being developed that take advantage of fine-grain control and construction at a very small scale.

New kinds of sensors, too tiny to see, will detect events that occur on a microscopic or submicroscopic level, in places that were heretofore impossible to penetrate. For instance, tiny sensor-bearing engines could be ingested to seek out the presence of cancer cells within a human body, from the inside, signaling tumors when they are still small enough to remove.

As I write this, a small Emeryville, California, company, Nanomix (in which I am an investor), has demonstrated the first monomolecular hydrogen sensor on a chip. Small-scale devices like these also make new kinds of industrial processes possible—constructing materials by adding one tiny layer after another, in the same way that a crustacean constructs it shell. Starting in five to ten years, in other words, human beings will have the ability to mass-produce new materials, in clean and efficient fashion, with minimal raw materials and very sophisticated design, by operating at the atomic and molecular levels. As these new industrial processes arrive online, we will see a variety of new kinds of materials come to market. Some will be sophisticated alloys of metals; others will be new kinds of polymers and plastics. Some will be complex interactions of biological and electromechanical systems with unique performance characteristics, able to change character, color, shape, texture, or form upon command.

For example, in 1998, researchers discovered carbon nanotubes—

filaments of graphite with unusual elasticity, strength, and control over electric conduction. These would allow computer displays to be built into walls, and buildings to snap back into shape after an earthquake. Nanotubes are a thousand times lighter and stronger than steel. One could imagine an airplane made out of them that could change shape during flight, to adapt to the distinct aerodynamic demands of takeoff, horizontal flight, and landing. There are two major constraints. The first is cost; the raw material from which carbon nanotubes are constructed is currently ten times as expensive as gold per pound.[61] And the second is installed base. From the experience of plastics and other synthetics, history tells us that new materials come to market very, very slowly.

Nanotechnology is one of several "stage-one" technologies that are predetermined to appear soon. We know they are coming because we can see them in the pipeline. Another such technology is more powerful and flexible computing. We have already seen in Chapter 4 (The Return of the Long Boom) how media infrastructure is poised, waiting for the development of the "last mile" of inexpensive broadband. Meanwhile, however, "Moore's Law" still holds; computer devices continue to become twice as powerful per dollar every eighteen to twenty-four months. Ancillary devices, such as storage disks and display screens, also continue to come down in price. Most computer users are so accustomed to this that there are very few surprises left.

In the near future speech recognition will be commonplace. Flexible-filament computer display screens will be printed on cloth at $40/yard and tacked up onto walls. Personal data assistants, cellular phones, wireless Internet receivers, and laptop computers will continue to converge into the universal pocket computer. People in business will routinely operate machines that would have been considered supercomputers just two decades ago—and still use them, as they did two decades ago, to run Microsoft Word.

The great battle still to be fought, in the United States at least, is over privacy; computers and sensors this powerful will inevitably be hooked together to monitor all forms of human activity. The recent arrest of musician Peter Townsend, after he used his credit card at a child pornography Web site, is a harbinger of the law enforcement of the future. Another harbinger is the advent of automated toll-booth devices on automobiles, such as E-ZPass in the Northeast, which can track the movements of cars and provide a record of your travel patterns—and issue summonses or cancel the devices when the driver speeds through a tollbooth.

In developing the film *Minority Report* we imagined a future in which privacy—the ability to hide activities from government monitoring or conceal identities—was virtually nonexistent. Access to any medium-security building involved retinal scans, which uniquely identified each individual without an ID card. The only way to defeat a retinal scan is to acquire a different retina through eyeball transplants—which became a key plot element in the film. In researching that technology we learned that the retinal scanner is less than a decade away. They'll be commonplace by 2012. A decade or two after that scanners on streets will routinely identify passersby via retinal matches.

In such a world there are no secrets. To many the inevitability of such a world will come as a terrible surprise. There will be many protests and efforts to restrain the technology, and in my opinion, they will fail. (That's not inevitable, but it's so likely that I have a hard time imagining any alternative.) The technology of intrusion will be so powerful, and its ability to penetrate secure systems so enormous, that it will be far more difficult to hide. Criminal organizations will continue to protect themselves; but only if they are willing to kill members who leak information. To escape from such a system will require extreme ingenuity. You would have to use only cash, which means you could not travel by airplane, train, or rented car. You might well have to establish a false identity, or move to a primitive society.

A final stage-two technology is the growing level of control over biological processes, particularly those involving aging, reproduction, and disease prevention. We saw in Chapter 2 that research on aging and other applications of genetic engineering have accelerated.

The same is true of fertility treatments, such as in vitro fertilization (IVF) and various types of drug treatments for infertility. These are available today, but they are also paving the way for a threshold leap, over the course of the next ten to fifteen years, in our understanding of the nature of plant, animal, and human biology, and the ability to intervene at a variety of levels.

Stage Two: New Frontiers of Pure Science

It used to take a generation to translate a discovery on the frontier of science into a technological change. In our time, in nearly every arena of science, this delay is getting shorter and shorter, nowhere more so than in the realm of biology. Pure research can yield new medical and biological advances in a few years because there is vast financial support and demand is so high for them. That will continue to be the case. The principal delay in finding new drugs and therapies is no longer mainly science. Rather it is increasingly the regulatory processes that stand between the laboratory and the doctor's office.

What, then, might we expect from the frontiers of biology? It is very likely that a large number of bioindustrial processes will be developed that enable us to grow new materials, fabricate chemicals, and even ultimately construct buildings. New bioindustrial complexes will produce pharmaceuticals, medicines, fibers, and foodstuffs. Their principal input may well be clean water, with all other raw materials grown onsite and manipulated through molecular engineering. Scenario planner Kees van der Heijden speaks of bypassing agriculture: growing steaks in steel vats, indistinguishable from the best-quality beef, and far less destructive to either animals or the environment.[62] The same science will be used for the regeneration and repair of human tissue.

This will require, to say the least, some getting used to. Probably the most dramatic change is to accept a much higher level of choice in biology. The ability to choose the design of our produce, our livestock, our bodies, and our children will expand, beyond the limits anyone might now imagine. Current "normal" (allopathic) medicine is based on the idea of eradicating disease, either through chemical interventions (poison what we don't like) or surgical interventions (cut out what we don't like). Preventative health care begins and ends with interventions in the behavior of the body's host (reduce what we don't like gradually). But regenerative medicine proposes that we can reprogram cells as if they were computer chips, to grow new tissue or replicate themselves in particular ways. Need a new kidney or a new heart? Rather than wait for a transplant, regenerative medicine suggests that you can grow it yourself. Craig Venter of Celera Genomics, who led the successful private effort to sequence the human genome, has begun to work on how build new species from scratch.

Hungarian biologist Karelyi Nikolich, for example, has discovered that some types of neural stem cells have regenerative ability. If they are injected into a stroke victim, they can migrate to the site of the stroke and repair the damage. No one yet knows why the cells go there, but they do, and they can grow new tissue. They don't regenerate the memories that were lost, but the ability to retain memory returns. The result is a vastly more functional brain and a much faster recovery time. (I am associated with Nikolich's company, which is called AGY—a spin-off of Genentech.) A Swiss company called Modex is growing skin for victims of burns and severe wounds, using their own tissue to do it.

An increasing share of biological research is either illegal or unpopular in the United States, because of religiously motivated opposition to the concept of cloning—and because of environmentalist concerns about genetic engineering. No matter how valid these arguments are, they are predetermined to fall on the losing side in the long run, simply because other nations will pick up the slack. In addition, the kinds of innovations that disgust and frighten people tend to become commonplace as they are put into use. I can remember the moral debate surrounding in vitro fertilization (IVF) in the 1970s. Louise Brown, the first "test-tube baby," was regarded as a kind of freak. Today, there are hundreds of thousands of people alive who were conceived through IVF. And there are very few prospective parents, if any, who would turn down IVF for moral or spiritual reasons if they could not conceive in other ways.

Cloning, for all the drama associated with Dolly the sheep, is not much of a leap from IVF. The main consequence of today's debate over cloning is its social impact; it will have prepared the intellectual and emotional ground for people to cope with the effects of much more sophisticated biological technologies. The general direction is clear: we are moving from a new understanding of basic biological principles to a much more fine-grained level of control over processes, down to the cellular and DNA levels, that were previously uncontrollable.

Something similar, albeit much less controversial, is going on in chemistry. Nanotechnology, a stage-one enterprise like genetic engineering and fertility treatment, also has some stage-two implications. Until recently, chemists operated on the aggregate-substance model; they mixed together large numbers of atoms and molecules to produce relatively large-scale aggregate chemical reactions. These reactions weren't always identical on an atom-by-atom level; but because there were so many atoms in a vat or a beaker, the statistical probability was high enough that the interaction would be predictable. Anomalies could be discarded. The mathematical field known as thermodynamics evolved to explain the behavior of these large-scale chemical reactions over time, and chemistry became highly predictable—so long as chemists restrained their research to large numbers of molecules and fairly coarse-grained procedures. Most of these reactions occurred at relatively high temperatures, such as above the boiling point of water.

We are now beginning to develop a chemistry that operates on the level of individual molecules, atoms, and bonds between atoms. These provide a much finer-grained set of explanations for why different substances combine in particular ways, and they offer much more particular forms of control over the substances themselves. We can now line up individual atoms to create unique "designer" molecules that would never be created (for example) in a high energy-consuming reaction. Can we design those molecules, and fabricate them to meet our industrial needs? The answer is possibly yes. Nature, of course, does exactly that in its biological systems; wood, seashell, spider silk, and fur are all examples of sophisticated chemical structures that are generated by living systems. As the science writer Janine Benyus suggests in her book *Biomimicry*, by emulating such processes we may be on the verge of our own industrial revolution.

Quantum Computers

The effects of stage two may be compounded by a significant research effort under way now: the quantum computer. The quantum computer's success is not predetermined; but if it *is* successful, it will spark an order-of-magnitude acceleration in the pace of research and development. In terms of sheer computing power the leap from today's computer processors to the quantum computer goes far beyond Moore's Law; it could be as revolutionary as the leap was from the single transistor to the microchip. Computers that are literally billions of times more powerful than today's computers could be applied to unbelievably hard mathematical problems: protein folding, gene design, mapping the universe, controlling very complex systems, modeling the climate, developing more complex forms of encryption, and possibly meeting the long-standing, never-before attainable goal of artificial intelligence.

I began to explore the potentials of quantum computing in detail in mid-2002, when the Defense Advanced Research Projects Agency asked me to help develop a strategy for it. For example, should the United States launch a "Manhattan Project" on quantum computing? Could we get a feasible prototype launched by 2010, if money were not an issue? The answer was a qualified "maybe": the early signs are good but the challenges remaining are very large. DARPA will be exploring prototypes over the next two or three years and investing more money if the technology shows promise. There is a possibility that the effort will never come to much: the U.S., after all, has poured probably $30 billion into researching fusion power during the last thirty years, with negligible results in the end. (In the 1960s fusion power was believed to be forty years away from commercial production. In the 1980s it was forty years away. Today it's *still* forty years away.) It's hard to tell whether quantum computing would be more like the microchip (which paid back its government investment many times over) or more like fusion power. But so far, the technology looks both plausible and extremely promising.

Conventional computers operate in ways that are completely consistent with Newtonian physics. Information is encoded in the electrons of highly conductive materials. But the electrons are part of many atoms that encode the binary bits, 0 or 1. As the signals from the electrons change, they combine in vast quantities to provide the computer's "output": first in binary code, and then in translations of that code into either software commands or data. The binary system, in short, sets the limits of the performance and capabilities of the computer.

Quantum computers still encode information in the behavior of electrons, but there are two other characteristics in the design that enable a far more complex form of computation, even at the electron level.[63]

First, electrons (and other quantum particles) store information not just as binary zeros and ones, but in effect are in both states simultaneously—or in "superposition," as it is known. Each electron can be encoded in a variety of states, and a different number can be extracted from each. The phenomenon of superposition allows an encoding scheme in power for data that grows faster than the complexity of the problems, so more complex tasks can be accomplished faster.

Second, two quantum particles can be "entangled." This means that once two particles have been connected to each other, they can be physically separated but still maintain a mutual and instantaneous influence, almost as if they retained a mystical link with each other. If the charge of one particle changes, the charge in the other particle will change as well. This sounds far-fetched in a variety of ways (among other things, it would break the limit of the speed of light as a constraint on the transfer of information), but it has been replicated in a number of laboratories, most recently Los Alamos, where quantum teleportation was achieved in the open air. Two photons were sent seventy yards between a transmitter and reciever. The charge in one photon was measured, and the charge in the other photon was determined accordingly. (This phenomenon in itself could revolutionize encryption; entangled photons could carry the keys for an encrypted message; they would be perfectly secure and distributed instantaneously.)

There are, of course, quite a few problems that must be solved before a reliable quantum computer can actually be built. For example, can the quantum states of the electrons persist long enough to be useful? Can data be reliably put into them and extracted from them? Can error correction be performed on them, so that their reliability can be tested? There already are theoretically based yes answers to the first two of these questions. Error correction is more complex; researchers can test the computations to three or four levels of operation, but tests to about a thousand levels are needed to guarantee reliability. The technology is thus a long way from certainty. Moreover, the solutions to these problems probably do not lie in conventional computer design. Quantum physics is innately weird and counterintuitive, and the solutions are likely to be weird and counterintuitive as well.

The effect of having computational power at this massive scale is almost unimaginable in advance; it would change the world dramatically, in ways that cannot be predicted. Even if we have a goal in mind, we still would not understand the limits that exist on our ability to reach that goal, from the sheer complexity involved. For example, to precisely manipulate DNA, a researcher must be able to understand the full complexity of relationships among the relevant genes, and how they interact. A particular physical characteristic, such as eye color or nose shape, is derived from the interaction among many different genes. We can manipulate genetic information at an aggregate level, but we don't even know if it is possible (for example) to change someone's eye color from blue to brown, or vice versa, through genetic manipulation, or what the consequences for other gene sequences and characteristics would be if we did.

Similarly, the vertical magnetic-levitation highways seen in *Minority Report*, in which cars zoom in and around each other in complex cities at all levels—on the street and in airspace—are probably not feasible without quantum computers. More precisely, without quantum computers we don't have the computational ability to determine whether they're feasible or not. Nor, without more powerful computers, can we gain a clear conception of the kinds of interventions that could be made to ameliorate storms and other forms of dangerous weather.

We *can* envision some of the complex systems in which quantum computing would make a difference right now. Protein folding, for example, is a long-standing problem in biological mathematics. Proteins, the primary building block of animal and plant tissue, assemble themselves into their cellular forms in a way that is somehow related to DNA sequences. If that relationship could be modeled and understood, it might help us understand a large variety of protein functions and malfunctions, including such diseases as cancer, Alzheimer's, cystic fibrosis, and HIV. The calculations required to model the long geometrical chain of protein formation are so complex that they exceed the capacity of all the world's supercomputers put together. While it has been possible to make some headway using networks of computers linked by the Internet (a process called distributed computing), the field will probably not take off as it might unless quantum computers (or some similar advance) become available.[64]

When I was an astronautical engineering student in the late 1960s, the turbulent flow of air over a wing could not be mathematically modeled. You could only study physical analogies in wind tunnels and similar tools and then design a wing accordingly. In the early 1990s, using relatively small computer chips linked together to comprise a massively parallel supercomputer, the supercomputer researcher/entrepreneur Danny Hillis modeled turbulent flow in terms of bundles of six atoms—a far more precise measure, but still not close enough to predict exactly how the wing would behave under all circumstances. Quantum computers, if they work, will be able to simulate and distinguish the flow of each atom over the wing precisely, and allow engineers to design the wing accordingly to meet a variety of circumstances.

Incidentally, if the quantum computer does *not* turn out to be viable, and no other computer technology emerges, then there is another inevitable surprise on the horizon: an end to Moore's Law. Somewhere between seven and fifteen years hence (i.e., between 2010 and 2018), computers will stop doubling in speed and power per dollar every eighteen months. These efficiency gains in computing power depend on shrinking the space between processors, and microprocessors are growing so small that they are close to reaching the molecular level already. Without quantum computers or some other technology based on the new physics, they will hit a limit beyond which they can't continue to shrink. That surprise would itself be a shock to the global system. It would remove the productivity driver for the "Long Boom" economy. And it would slow much scientific and research advancement.

But if quantum computing turns out to be viable, then the typical computer of 2020 could be a hundred million times as powerful as the computer of 2003. Even then, most people will still be using them to run word-processing programs. Will the Microsoft Word of 2020 resemble the Word of 2003? Or will it have some new kind of relationship with the thinking process that is barely conceivable now? Will quantum computing accelerate our ability to write programming code? And to publish innovative new software? And will the software produced in this future resemble the software of today, or will it have evolved into bold new forms that take advantage of these powerful new machines?

Many uncertainties remain.

Phase Three:
Energy, Reality, and Space

There are three areas of long-term scientific-frontier speculation that I am paying the most attention to these days. The first, as I noted at the front of this chapter, is dark energy. Suppose that there is, in fact, some cosmic force that counterbalances gravity, and causes the universe to expand at an accelerated rate. Then suppose that we could tap that dark energy in some way—at minimal cost, with no environmental effect. Suddenly, the concept of energy technology changes completely, in a way that wasn't imaginable before. There could suddenly be a little device that sucks the energy out of space, just as a radio or television receiver sucks sounds or images out of space, via electromagnetic radiation.

It's hardly an inevitability. But I would not be surprised if someone started to develop both the science and technology of dark energy, in a way that lies entirely outside the space of current solutions.

The second area I am paying attention to is the application of information theory to understand the underpinnings of reality. The most significant researcher in this field is Stephen Wolfram, whose self-published book *A New Kind of Science* is an eight hundred-page manifesto about the nature of reality. Wolfram, a physicist and inventor of the *Mathematica* software program, has studied "cellular automata" in depth. These are computer simulations that typically begin with patterns of colored pixels on a screen, and apply simple abstract rules to those pixels. Wolfram identified 256 possible simple rules for black and white pixels; they each represent a different variation on the theme, "Change color if the pixels around you match the following pattern. Don't change color if they don't." Then those rules are applied to a group of thousands of pixels, time after time after time after time, as only a computer can apply them. And strange, intricate, beautiful, and highly complex patterns start to appear on the screen, with recurring features. Of course, the number of possible rules expands dramatically when you add more colors besides black and white; and if you could extrapolate the number of pixels to match, say, the number of atoms in the universe, and you could expand the number of rules to reflect the number of "colors"/ dimensions in real life, you might get something approaching the nature of reality as it exists today.

When he applied different rules to his patterns of pixels on computer simulations and let them "run" for many iterations, Wolfram observed that many of these patterns simply "died out"—they collapsed into some stable, simple result. But others grew more and more complex and intricate. The rules governing these more complex systems were not necessarily complex in themselves; simple rules could sometimes produce very complex results. Wolfram began to track the recurring effects of different types of rules, and came to the conclusion that reality itself was operating on some of the same basic principles. Very simple rules, applied to simple components like atoms and cells of living beings, had produced the remarkable and increasing complexity of our world. Such complex phenomena as human vision, pigmentation patterns in animal skin, the growth of crystals and snowflakes, and the random pattern of fractures when a piece of glass is broken, can all be modeled with the rules-and-iterations approach of Wolfram's mathematics.

If this view of the world is as replicable and generative as Wolfram suggests, then it represents nothing less than the transformation of physics into information theory. There is a code, in other words, that determines what happens when atoms and other small entities come together, and it is possible to crack that code.

Meanwhile, other physicists are portraying the ultimate structure of the universe as comprised of a great variety of different kinds of submicroscopic fields: very small ranges of energy or attraction in which all the entities interact and affect each other. The interaction of these fields in sophisticated ways becomes tangible as vibrations, which then combine into particles like electrons or protons. These, in turn, generate everything we think of as energy and mass in the universe.

This field theory of matter is conceptually very rich, and it may become part of a significant new way of looking at submicroscopic reality. But it doesn't explain how a particular pattern of vibrations becomes a particular electron or proton. Knowing the alphabet does not give us Shakespeare. If Wolfram is to be believed, that causal relationship can be understood through information theory. Different submicroscopic fields, following a set of simple information "rules," come together to produce patterns of great complexity. It is as if there were something called the "reality code," which generates the principles of physics, chemistry, and biology: why particular particles create particular atoms in particular combinations to produce particular substances and organisms.

Reality, in short, is a giant computer, and it conceivably could be programmed if we knew how to input the right "data." We may once more be on the verge of a fundamental redefinition of physics, akin to the quantum and relativity revolutions. It may even be possible that this new redefinition will resolve some of the inherent ambiguities and abstractions of quantum theory and relativity. Wave particle duality, curved space, and quantum uncertainty were all verifiable through instruments, but they were impossible to see and hard to imagine in any concrete way. What does curved space "look" like? No one can say. You can't even see it through an instrument, like a microscope or spectrometer. But Wolfram-style physics, with its use of simulations and rule behavior, may give us a leap toward a more easily comprehensible, and even visualizable, structure of the universe.

What kinds of technologies would be possible if we understood the code? Perhaps nothing of significance. This is such a radical idea that, even if it is correct, it may not be "implementable" in any technological sense. It may not be possible to construct instruments to measure, or devices to manipulate, the reality code of the universe.

On the other hand, this theory could lead to such innovations as real-world teleportation. Sending matter from one end of the earth to the other (including sentient matter, like human beings) may be as simple as sending the right code. *Duplicating* matter, similarly, might be as simple as duplicating code. If these two technologies are in place by 2050, then they eliminate all problems of traffic, air pollution, or famine. Travel as we know it has changed; people instantly appear in the place they want to go. Manufacturing as we know it has changed; people instantly duplicate anything they want a copy of, including unlimited quantities of food. Of course, a whole raft of new problems has emerged to take their place.

Quantum entanglement may also foreshadow teleportation. Already there are successful experiments in which the physical states of entangled particles like photons have been sent the length of Lake Geneva, between Swiss laboratories. There's still a long way to go from such rudimentary experiments to get to the George Langelaan short story "The Fly," let alone to "Beam me up, Scotty." But one can imagine crossing the transporter divide during the next fifty years.

The other area I follow closely is space. The short-term news here is discouraging. Even before the tragic astronaut deaths on the *Columbia* on February 2, 2003, the space shuttle program was an ill-fated venture. It was so inefficient and poorly conceived that each mission cost more than if we had simply thrown away the rockets each time we launched them. It also set a terrible precedent for space travel: expeditions that risked human lives for very little pure research or technological development. Ironically, the space shuttle represented an effort to play it safe—in the bureaucratic sense. NASA has done a very poor job of engaging the American people. Americans love the romantic idea of continued space travel, and are willing to invest in it; but instead of capitalizing on that popularity, since the Apollo missions NASA's leaders have played the technocratic, cost-cutting, budget-justifying game. These failings, compounded by the Columbia disaster and other budget priorities, may lead to a major retrenchment in the manned space program or, in the extreme, even to its being shut down altogether.

Sooner or later, however, space travel will inevitably enter a renaissance. It's not clear yet when that will happen. The most limiting factor is not the danger, but the sheer cost of the propulsion necessary to move people and materials outside of Earth's atmosphere. Unfortunately, recent research in rocketry suggests that improving on this may be more difficult than was previously believed. Many space enthusiasts had pinned their hopes on a form of aircraft called the "supersonic combustion ramjet," which could travel at eight thousand miles per hour to launch a vehicle into orbit. This has proved harder than anticipated; it turns out to be very difficult to keep the fuel and air mixture flowing in the right proportion at very high speeds without blowing out the flame. This problem may not even be solvable through the calculation capabilities of quantum computers; the difficulty has to do with the size of the drops of fuel, the unusual circumstance of traveling at high speed, and the innate physics of combustion. We can figure out how to leave Earth reliably only if we can figure out how to "keep the fires burning" in this high-tech venue.

However, as the Long Boom comes to pass, and as more people live longer, there will be a growing number of people who will want the remarkable experience of seeing Earth from space before they die. It is plausible that by, say, 2030, there will be tourist facilities in Earth's orbit or on the moon, where the very wealthy can take brief holidays, experience zero gravity, and be able to say that they are members of the very elite group of people who have actually left the Earth.

It's possible, of course, that other ways to escape Earth's atmosphere may be discovered, perhaps through the use of "dark energy," teleportation, or a new understanding of gravity. Even then, the challenges of travel and settlement within the solar system are immense. For instance, several writers have made a passionate case for terraforming and colonizing other planets, starting with Mars—rendering them habitable for human, animal, and plant life and then building settlements there. (This is presumably possible, but it's a *lot* of work. It could take as long as a thousand years.) It may well be that by 2050, the Chinese, Europeans, and Indians will have taken up the challenge of space travel together. A mission to Mars could be a great unifying factor, drawing all of the orderly nations of the world into a massive multilateral project.

And if the physics has advanced enough, then we may be able to dream even grander dreams than colonizing Mars. Astrophysicists have abandoned the premise of star flight, because of the inescapable constraint of the speed of light. Current physics tells us no faster speed is possible; thus it would take far too long to reach even the closest solar system. A new physics paradigm might just possibly tell us that faster speeds *are* possible and allow us to plausibly conceive of (and develop means of) travel to other stars. No such physics paradigm is on the horizon right now; but the most important inevitable surprise about pure science can be summed up in four words: The horizon always shifts.

CHAPTER 8

A Cleaner, Deadlier World

There are two kinds of conventional wisdom about the natural environment today, and both of them are wrong. More precisely, they're both partly right. On one side many people believe that our planetary natural environment is poised on the brink of crisis, the kind that could devastate civilization. Toxins, they argue, are building up in human and animal tissue, they're responsible for unprecedented levels of cancer and infertility. The ozone layer is still being depleted; old-growth forests and wildlife are disappearing, aquifers are being depleted, and carbon dioxide and other greenhouse gases continue to accumulate in the atmosphere, with tempestuous climate change about to hit crisis-level proportions as a result.

On the other side people argue that there is still no proof that human beings are causing "global warming" or any other sort of climate change—and that geological evidence suggests that the earth has gone through multitudes of climatic fluctuations in its 5 billion years of existence. They say that the environmental threat has been blown out of proportion by special interest groups with an overt antibusiness political agenda, and that with enough attention to business as usual and economic growth, the ecological future will take care of itself.

Advocates on either side have plenty of evidence to point to, much of it inconclusive. Fortunately for the rest of us, it *is* possible to get a clear sense of the truth in this arena—surprisingly possible, given the ambiguities of most forms of scientific measurement and modeling of complex systems (and if there ever was a complex system worth modeling, it is the interplay of human industrial activity and our environment). While there are no big global certainties, in the sense that a single computer model or instrument can predict the future environmental quality for the planet as a whole (or even for any region), there are lots of little certainties. We know how to measure the incidence of pollutants, temperatures, water availability, land use, soil quality, and quality of life; and we know how to put that data together comprehensively, so that our whole understanding is a lot richer than the sum of the parts. We can actually say, with some certainty, not just what has happened, but what is going to happen.

Much of the news is good—surprisingly good. To most of us who are accustomed to constant forecasts of environmental disaster, one of the major inevitable surprises is the need to get used to good news. There is a growing body of evidence that the biosphere is becoming healthier every year. Pollution is diminishing. Species are harder to extinguish than they seem, and some are actually returning from the brink of extinction. Wilderness lands are being reclaimed from farming. The environmental challenges facing us are huge, to be sure; but we've already risen to meet many of them. We are not going to experience the *Waterworld* or *Blade Runner* style environmental crisis that some of us have anticipated for more than twenty years now.

But the complacent should not rest easy; because in a few specific ways the assumptions they hold about growth and health in the future are about to be inevitably shattered.

There are in fact three types of environmental and health crises on the horizon. First, the expected calamities that won't, in fact, materialize—but will nonetheless be nervously anticipated. These unrealized calamities will continue to influence the world because of the measures that people will take to avoid or counter them.

Second, we can expect inescapable environmental calamities related to global climate change. We know that they are coming, and we even know when—because they have already been with us for some time. These calamities are changing the world, in part because of the ways in which human institutions have already begun to adjust to them, and in part because of the limits they place on human endeavor.

Finally, there are the truly worrisome calamities. We know they are coming, but we don't know when. We merely know that they have not (by and large) been prepared for. Leaders, institutions, and the public at large are all failing to anticipate them—thus ensuring that, when they come, their impact will be more devastating than anyone currently expects.

The Population Bomb Fizzles

In the first category of expected crises that *won't* materialize, the biggest surprise is population growth. As mentioned in Chapter 2, it's about to level off. Humanity passed the critical inflection point for population growth three decades ago; in the 1960s the rate of global increase began to slow down. It's taken until now to see many of the effects. It's a bit like slowing down a car without using the brake; just as there's a time delay between the moment you lighten the accelerator pressure to begin coasting, and the moment the car starts slowing down, there has been a delay between the lessening of population growth rates and the stabilization of the numbers of people on Earth. But that number is stabilizing, and in some wealthy regions, like Europe, it is actually decreasing. The inevitable implications have not yet dawned on many people, particularly those who once expected humanity to overrun the planet.

There are currently about 6 billion people on the planet Earth. Twenty-five years ago there were 4 billion. This was an unprecedented example of population growth. Many demographers expected the population to continue doubling faster and faster (or at least at the same pace), reaching as high as 25 billion people by the middle of the twenty-first century. After all, more children born means more women growing up, which means more mothers coming of age, which means more children born . . . doesn't it? This was the dire fate that Stanford biologist Paul Ehrlich called "the population bomb" in the 1960s. There were serious worries about whether the Earth's carrying capacity could support that many people, or whether, like passengers on a boat with a limited room for provisions, we would have to jettison some people or watch them starve.

Then the acceleration decreased. At first, in the early 1980s, many demographers barely noticed this, or dismissed it as a fluke. Then the data would no longer let them dismiss it. Gradually, they began to change their predictions, saying that the human race would only rise to 15 billion people before it capped off. Then 12. Then 10. Now, they are projecting a rise to 9 billion.

To be sure, 9 billion is still a lot of people. But it is a far more manageable number than 25. And the expected number is still falling. Demographer Chris Ertel puts it this way: "The last doubling of the species on Earth has already happened. We are not going to double our population size again." In fact, the demographic decline keeps taking place faster than expected.

The underlying cause is simple. Birth rates are a function of the number of children that each adult woman can produce. For every woman who bears her first child later in life, the population growth rate falls a little. And as women gain control over reproduction, they naturally postpone bearing children. A first-time mother over age thirty-five used to be an extremely rare individual; now, in places like Japan and parts of Europe and the U.S., there are thousands of them. And those are the places where the birthrate is falling most dramatically.

Some demographers have assumed that the leveling-off only occurs in wealthy nations, and that impoverished developing nations will continue to have a population boom—but it's even happening in many of the poorest nations in the world. In the summer of 2002 new demographic figures were released for East Africa, one of the poorest and most troubled regions in the world. The birthrate had begun to fall more quickly than anticipated. No doubt civil and regional conflicts and the spread of AIDS were factors; but only some of several. Even amid economic or health crises women are choosing to postpone the year of their first childbirth.

There were three places in particular where the population bomb once seemed most threatening: China, India, and the Middle East. But two of these three hot spots have been defused. China no longer has a "replacement" birthrate—its new babies are no longer numerous enough to replace the adults who die. This occurred largely through China's famous "one baby" policy, which I dealt with in Chapter 3. Even though the "one baby" policy is no longer in place, it still casts its demographic shadow over the Chinese birthrate. Meanwhile, in the wake of tremendous economic growth in the 1980s and 1990s, India has also moved far below its projected population growth. Only the third instance, the Muslim nations of the Middle East, are experiencing the kind of population growth forecast for them in the 1980s—and even there the trends point toward deceleration.

Wealth and Turnover

There is more good news in the industrialized world. When he first proposed the population-bomb idea, Paul Ehrlich offered a corollary: that environmental degradation was a direct function of affluence and technology. He wrote the principle down as a formula: E = P • A • T. (Environmental Impact is equal to the product of Population, Affluence, and Technology.) Ehrlich argued that the wealthier people became, the more goods and services they would consume, their technology would wreak greater environmental devastation.

We now know that he was wrong, at least in the long run. To be sure, in the short run, as people become more affluent they consume more: they drive, heat homes, eat food from disposable containers, and do all of the other energy- and materials-profligate things that we associate with a consumer-culture lifestyle. But over time the relationship between affluence, technology, and environmental quality is far more complex. Technology and affluence have led to dramatic *increases* in environmental quality around the world, sometimes in unanticipated ways.

Moreover, those effects tend to naturally emerge in the highest-tech, most affluent societies. The richer a society is—the more wealth per capita—the cleaner it tends to be. Rich and clean go together. The worst environmental decay of our time is taking place in poor countries, not rich ones. Air and water pollution in most rich countries is far better today than it has been in centuries; without coal-burning and the dumping of human and animal waste in the streets, for instance, the smog that hovered over London (the notorious "fog") from at least the thirteenth century is now a thing of the past.[65] "Particle pollution in London," reports researcher Bjorn Lomborg, "has decreased twenty-two-fold since the late nineteenth century."[66] In rich countries old-growth forests and other forms of wildlife are restoring themselves, or in some cases being restored, as farmland recedes. Scandinavia and Switzerland are leaders here, but hardly unique. The United States, despite its reputation as an environmentally careless country, has an enviable record of restoration: in the last fifteen years thousands of acres of old-growth forest have been replanted.

Some of the reasons for the link between environmental quality and per-capita wealth are well known. Wealthy and middle-class people have the time and political clout to demand more environmental quality from their governments, and enough education to be attuned to its value and skeptical of the official excuses they receive. They tend to be more educated, which means they have a raised awareness of the significance of environmental quality. They also have a natural interest, often parodied as "NIMBY" (Not in My Back Yard) politics, in protecting their investments in housing and land. They have the time and wherewithal to take an interest in preventing extinction in plant and animal species (although the loss of species is still a grave trend, and one for which there is no easy overall solution).

Most important, wherever there are wealthy and middle-class people, there are businesses with the capital that is needed to invest in new, cleaner technologies. The form of economic growth known as "sustainable" development (development that improves the quality of life for successive generations, instead of reducing it), depends upon the innovative and wise use of technology. Over the past twenty years, on both the production and consumption ends, technology has quietly but steadily improved enough to make the concept of sustainable development viable. For example, air quality in most American cities today is much better than it was twenty years ago, because the automobiles of today are 95 percent cleaner than they were twenty years ago. Each generation of refrigerator, television set, plastic package, washing machine, battery, elevator, machine tool, oil refinery—it doesn't matter what form of technology—is a little more efficient, less wasteful, and less polluting than the generation before. New windows installed today have higher "R-values" (a measure of energy efficiency) than new windows installed five years ago. This long-term trend has been going on for at least a century. Every time engineers remodel a device, they make it more efficient than the previous one. Some feel obliged to do this for the environment's sake, but for most it's simply part of the continual effort to outengineer the competition's devices.

In *Minority Report* one plot twist makes the lead character—the police officer John Anderton, played by Tom Cruise—dependent on a tenement refrigerator. On the outside the refrigerator looks practically the same as one that was sold in, say, 1955. In specifying that look we reasoned that the "look and feel" of inexpensive refrigerators is unlikely to change much; they're usually purchased by landlords or people without much money, and design isn't a premium. But the inside is another matter. *Minority Report* is set in the year 2049, which means that refrigerator would probably have been manufactured in 2015 or 2020. That would make it an antediluvian energy hog and pollution machine by 2049 standards, but its compression and freezing technology would still be orders of magnitude cleaner and more efficient than even the most advanced refrigerator available for sale in 2003. The average refrigerator of 2003, meanwhile, looks just like its counterpart of 1973—but it is less expensive, and uses 80 percent less energy.

In effect, every industrial society enters a race between the rising use of materials and energy and the growing efficiency and effectiveness of its technology. Over time efficiency and effectiveness tend to win; they increase with more steady, predictable force than the use of goods and services, which rises rapidly during the initial stages of industrialization in a society, but then hits a plateau as the population levels off and people accumulate enough goods. There comes a moment, after all, when there just isn't that much excitement in the chance to buy a washing machine or toaster-oven; when goods become a burden, and people decide they would rather spend their money on cultivating gardens, or even giving it away. At that moment society as a whole takes a definite tilt toward greenness.

This is very good news. If Paul Ehrlich's formula *were* true, and affluence was a direct cause of environmental destruction, then we would be heading into a very problematic time. The energy, transportation, and water needs of the new world are likely to dwarf anything ever seen before in human civilization. For example, China currently requires a new thousand-megawatt power plant every single month to keep up with 12 percent growth in the demand for energy every year. It's a daunting challenge, no matter how those plants are powered; but if they had to be powered by current coal-burning technologies, with their impact on air pollution, then China (and all of Asia) would be in a terrible mess. Worldwide demand for energy and products may triple during the next few decades. It may grow even faster.

How rapidly, then, could a society like China—or, for that matter, the United States—grow "green" by cutting back on its current pollution levels? The critical factor is not the pace of innovation, but the willingness and capability of its people to turn over old capital stock. Even if new products are being refined and reinvented at a breakneck pace, their impact is still determined by the bottleneck of last-generation glut: the speed with which society can retire its old, highly polluting machines from day-to-day use. Eighty percent of the smog from internal combustion engines today comes from the oldest 20 percent of the cars and trucks on the road. The same is true for turbines, power generation equipment, motors (especially motors), and all forms of transportation. Air and water quality are dragged down by used equipment, bought secondhand and operated without warranties or manufacturers' maintenance schedules. Paradoxically, the more people throw away their old goods (particularly appliances and other energy-using goods), the cleaner the environment becomes.

This is the logic behind environmental scientist Amory Lovins's "negawatts" concept; the energy savings from turning over old, inefficient industrial infrastructure can actually be treated as a supply of new energy. In principle it would be a good idea to speed up the turnover rate by any means possible; but in practice this raises a variety of political and economic dilemmas that have yet to be resolved. Suppose, for instance, that the government of California offers a purchase subsidy plan to take the oldest 10 percent of all vehicles owned by low-income families off the road. This could be seen (and perhaps rightfully so) as a tax imposed on middle-income people to subsidize new cars for the poorest. At the same time it could be seen as an attempt to punish poor people by forcing them to give up their $500 used automobiles, which are the only transportation they can afford. Inevitably, people will begin gaming the system—selling their auto credits, or temporarily reducing their income just long enough to qualify for a new-car subsidy. And what would happen when an individual who owned a vehicle at the 11-percent metric decided to sue the government, based on auto discrimination? (California did, in fact, attempt such a program several years ago, but it was rapidly withdrawn.)

Moreover, a rapid replacement rate doesn't *always* yield environmental gains. Those who retire old sedans may choose to replace them, not with energy-efficient versions of the same vehicles, but with pickup trucks and sports utility vehicles. That's exactly what happened in Europe, according to energy analyst Lee Schipper; after years of energy taxes and auto-efficiency initiatives, planners expected per-capita automobile fuel use to diminish in the 1990s. And it did; but it diminished twice as quickly in the United States, because Europeans began upgrading to larger, heavier vehicles. In the U.S., where larger, heavier vehicles were already the norm, new fuel-saving technologies made more of a difference.

The greening of industrial society cannot be prodded to happen faster—at least not easily. But it also cannot be stopped. There are many leaders of the Republican party in the U.S., for instance, who have dismissed environmental concerns or deliberately tried to undermine them. But their own constituents have forced them to think differently. When you get right down to it, most people—liberal *and* conservative—want to live in an environment that feels healthy to them. (George W. Bush's proposal for $1.2 billion in hydrogen-power research is one of the most prominent examples.) Nor is this just a matter of lip service or "greenwashing." Industrial companies are being swayed not just by the threat of negative public opinion, but many other factors: customer choice (customers will increasingly find it easier and cheaper to switch to cleaner energy sources), technological change (which makes cleaner systems more profitable, if only because they are more efficiently designed), the growing awareness of their impact on global climate change, and their own values and interests.

One can already hear the difference in the way that business leaders talk, particularly those involved with power generation. Starting about ten years ago, energy analysts and planners began to take environmental impact scenarios seriously; today, at their conferences and meetings, they talk about almost nothing else when making investment plans for the future. Meanwhile, all three of the largest energy companies—Royal Dutch/Shell, Exxon Mobil, and British Petroleum—have announced their intention to lead the transition away from hydrocarbon (fossil-fuel)-based energy supplies. They are aware that, if the environmental impact of fossil fuel use remains high, it will be replaced by alternatives. BP went so far as to rechristen itself as a company "Beyond Petroleum."

I was present one night at a dinner party with Mikhail Gorbachev, the last premier of the Soviet Union, and Randy Hayes, founder of Earth Day and the Rainforest Alliance. Hayes said, "Don't you think we need to control the multinational corporations? They're the ones destroying the environment of the Earth."

"We in the Soviet Union were the worst polluters on the planet," Gorbachev replied, "and we had no help from the multinationals."

The state-driven economies of communism have been the dirtiest offenders, by far, because they were unconstrained either by competition for efficiency or by public pressure for environmental health. They didn't have to follow the rules. The rest of the world, including the largest corporations, are learning from the communist example—and from the capitalist example as well. Corporate leaders may cut corners and resist regulation, but they are also eager to avoid presiding over the next Bhopal, Valdez, Brent Spar, or Hudson River PCB incident.

For all these reasons environmental quality is predetermined to rise in the wealthy countries of the world. Even in the worst-case scenario there will be a decade or two more in which population and affluence outpace technology. But then the Earth, and the lovers of nature and health, will prevail. And when you consider the technological shifts about to emerge in this arena, the optimistic scenarios seem more and more likely to come true.

Energy Technology: Making the Transition

Here is the inevitable surprise for energy technology: Patience is finally paying off. For twenty-five years or more environmental activists have been arguing that renewable energy sources—wind, solar, biomass, and hydrogen-based fuel—can and should replace the fossil fuel economy. Progress, especially in disseminating the new technologies, has been so slow that many people have given up hope. In the next twenty years it's finally going to happen. We are finally crossing the energy transformation threshold.

That doesn't mean the transition will be smooth. We can expect many energy price spikes, with immense and devastating short-term consequences, like the electricity price spike that struck California in 2001. Only part of the blame can be assigned to the manipulations of deregulating energy markets by traders like Enron; the spike itself was a symptom of the underlying realities of energy economics. The industrialized world enjoyed relatively stable, low fuel prices for almost twenty years, between 1984 and 2001; these stemmed from our own improvements in energy conservation (which helped keep fuel prices low) and the general state of peace (which precluded any politically driven oil price spikes like those of 1973 and 1979).

Unfortunately, after a while, low prices create tight markets. They repel investors. For the past fifteen years there has been relatively reduced U.S. investment in energy supply—including oil, natural gas, renewables, and nuclear technology—and a relatively high but steady growth in demand. Sooner or later came the inevitable result: intermittent gaps in supply (symbolized, you might say, by California's lack of enough power plants within the state). It takes several years to bring a major new power plant or oil drilling platform on-line, which means the lag time feels like an unbearable crisis. But as prices go up, new investors and producers are drawn in by the opportunity for high profits to fill the gap. Gradually, suppliers proliferate, the shortage becomes a glut, and the price is driven back down. This cycle is a classic story in commodity industries, and we are now right in the thick of it.

People don't generally like boom-and-bust cycles, but this one basically represents good news—even despite the short-term pain. Because this time energy producers of all sorts have been attracted back to innovation in a way that hasn't been seen since the end of the nineteenth century: with a multiplicity of technologies coming on-line simultaneously in every market. Around 1900 steam, electricity, and gasoline all competed to be the fuel that would drive the new transportation machine called the automobile, while coal, oil, and hydro power (dams) competed to be the source of the new infrastructure called "electrification." In 1902 petroleum oil had less than 20 percent market share as a fuel for motor vehicles; steam and electricity had 40 percent each. If you'd had to predict which would ultimately dominate the market, it's very unlikely that you'd have chosen petroleum oil; it wasn't widely distributed or accessible—hardly inevitable by any means.

Now consider the automobile of 2015. What will be its predominant fuel source? Will it be an all-electric car, plugged into an electric outlet all night to recharge? (The few prototypes of such cars already on the road are fiercely beloved by those who own them.)[67] Or will it be a "hybrid" combination of the internal combustion engine and electric power, like Toyota's Prius and Honda's Insight? (These operate by electric battery at low speeds; at high speeds the internal combustion engine kicks in to drive the car and recharge the battery using gasoline.) Will the automobile be powered by stand-alone fuel cells, such as a hydrogen-based power source, supported with a network of hydrogen-oriented service stations? Will there be turbo-generator cars on the market, using natural-gas turbines to produce electricity onboard? Will cars sport photovoltaic cells to channel sunlight into electricity, either in combination with another power source or purely through solar power? Or will gasoline remain the dominant fuel, in new types of engines that get far more miles per gallon and reduce pollutants still further? (This last option would allow the industry to keep its basic distribution infrastructure and engine technology.)

Nobody knows for sure. All of these technologies are plausible winners, even solar power (though it's hard to imagine a photovoltaic cell being developed that could propel an automobile from the tiny sunlight-capturing area of its rooftop). We don't know which technology will predominate because that depends on the vagaries of both technological development and business competition.

Fuel cells are very likely candidates. These are stand-alone devices that convert substances such as hydrogen gas into power electrochemically, using a membrane of catalytic chemicals to split protons from electrons. They have no moving parts and are almost noiseless. When hydrogen is the fuel, the cells only generate two byproducts: heat (which can often be captured and reused) and water. Until now fuel-cell use has been held back primarily by the weight and cost of existing devices. Both are dropping steadily, but will they drop quickly enough to be competitive with other sources? It's hard to say.

Ultimately, fuel cells will probably be developed that feed on hydrogen through the same kind of tank-and-pump infrastructure that internal combustion engines use to feed on petroleum gasoline today. The transitional path to this future would probably begin with the facilities of existing oil companies; in the United States there are already at least nine major refining centers where hydrogen is a by-product of oil and gas production. Currently it is sold and pumped to chemical producers; it would not be much of a stretch to develop a hydrogen filling-station structure from there. Indeed, the first hydrogen filling station in the San Francisco Bay

Area recently opened, near the Chevron refinery in Richmond, a small city north of Berkeley. It is intended for use with a new series of buses that will run on hydrogen fuel in the East Bay.

If this kind of infrastructure build-out continues, then fuel-cell-driven vehicles could eventually make up 30–60 percent of the worldwide automobile and truck population. It's very plausible, but hardly an inevitability. There may, for example, turn out to be problems with weight or explosion risk that render fuel cells impractical for moving vehicles. Or another technology may turn out to be easier to implement and market.

But whatever the ultimate fuel source becomes, we can be certain that it will be green. As competitors, the developers of all these technologies will push each other toward efficiency and emission reduction. Inevitably, the new cars of 2020 will emit far fewer pollutants than their counterparts today (perhaps even 95 percent less) and require far less fuel to drive. One study, coauthored by two automotive-design experts at Argonne National Laboratory and the University of Michigan respectively, foresees suburban-utility vehicles that average forty miles per gallon, almost double their average today.[68] Electronically controlled power trains, more efficient fuel-injection systems, "continuously variable transmissions" (with an infinite number of settings instead of four or five), electromechanical valve governors (with much finer-grained controls than the camshafts they would replace), faster starter motors (so that engines can shut down more often and idle less), lighter auto bodies, and more aerodynamic design will all play a part.

Another inevitable technological and environmental revolution is coming in power generation. Perhaps by 2020 and certainly by 2030, we will see a great variety of power-plant technologies competing to be the cleanest and most efficient on the planet. The movement toward cleaner, smaller-scale technologies—fuel cells, turbogenerators, and miniplants—for providing electric power to buildings is already under way. We will also see methanol-powered fuel cells (which are currently safer than hydrogen-powered devices) in portable electronics: laptops, PDAs, and cellular phones. As a part-time venture capitalist I'm on the board of one company, Neah Power Systems, which is developing this technology. In December 2002 I received the first fuel-cell-powered mobile phone call ever made. Like semiconductors the devices can be produced entirely out of silicon. Whoever ends up as the lead manufacturer, the device is likely to be popular for travelers; airport stores will sell the fuel as capsules, about the size of a disposable lighter (and as safe to use); we'll pop them into our laptop and replace them when the fuel is exhausted, every twelve hours or so.

Since they'll generate far more power than batteries, fuel cells will also make more powerful portable devices possible, including more powerful toys. They will also be significant military tools; the average American soldier going into battle in Afghanistan in 2001 carried 30 pounds of batteries, which represented less than a one-week supply. Fuel cells could replace all that. (The Defense Advanced Research Projects Agency, which coordinates U.S. technological research for the military, is beginning a massive new fuel-cell initiative.) A few years later we'll start to see larger-scale stationary fuel-cell power generation, first for isolated buildings and ultimately wired together in an electric-power counterpart to the wired-together computers of the Internet.

It's not inevitable, but highly likely, that another component of the power-generation revolution will be new solar-power devices—less efficient but much less expensive than currently available solar cells. The new cells cost $.045 per kilowatt hour instead of $.20–$.25. If you are covering your roof with solar panels, the new cells won't be durable enough for you, and they won't produce enough power to justify the trouble of installing and maintaining them. But if you're a power utility, you can place ten acres of the new cells in an arid field, and enter that power into the electric grid for a city like Mexico City or Houston. More northerly cities, like San Francisco and New York, are already moving in the same direction with wind power; others will refine the efficiencies of their hydropower generation. Ultimately, each major world city will probably have its own form of renewable "energy farms."

Fossil fuels—oil and natural gas, in particular, but also coal—will continue as fuel into the foreseeable future. But two clear concerns must be addressed in any discussion of them.

The first is availability: Will we have enough?

In the last few years it's become clear: The answer is yes. Even if war spreads in the Middle East or there is an extended crisis in Venezuela; even if the Caspian Sea regions and Russia remain too corrupt and remote to permit most oil companies to operate there; even though some notable industry experts have predicted shortages as early as 2004; we will not run out of oil. The current price spike may last a year or two, but not more. OPEC has already increased production to keep the price from getting too high. They rightfully fear that a high oil price will attract more competition from other oil fields and technologies. And these technologies are coming anyway.

I gained a visceral sense of the changes in oil exploration technology when I visited Shayba, a spectacularly large field of high-quality crude oil in a region of Saudi Arabia called Rub'al Khali, or "empty quarter." That is one of the most apt place names I've ever encountered. For almost a thousand miles in every direction there is nothing but sand dunes. Summertime temperatures range above 135°F. Temperatures in the "cool" season of November might drop to 97°F. The ground surface is hardpan salt flats, like the floor of the Black Rock Desert in Nevada. Piled on top are immense, powdery red sand dunes, hundreds of feet high, a spectacular wind-sculpted landscape.

The oil field was discovered in the 1960s, but it could not be developed until now; it requires a technique called horizontal drilling, to reach from a relatively accessible part of the desert to the buried, inaccessible oil. There is now a small industrial city located in a canyon where the hardpan floor is relatively accessible. It has hundreds of well sites, an oil and gas processing facility, shelter for the people who live there, and horizontal drilling cables fanning out in all directions through the ground under the dunes. If you went up the canyon walls and wandered a few hundred yards away over the dunes, you wouldn't see any of it. And you might never even be able to find your way back.

Deep offshore drilling technologies are similarly available to take advantage of oil in formerly inaccessible ocean sites. Nobody had even bothered to look there before, because no one could imagine drilling there. But now we can look, and we are finding vast reserves there as well. Ultimately, there *is* a limit to the amount of oil on the Earth, but—contrary to many predictions—we haven't come close to reaching it yet. And we won't for at least several decades, and probably much more.

The second concern has to do with environmental impact, and here the answer is not yet clear. Oil, natural gas, and other fossil fuels will indeed be limited by how hard it is relatively to reduce their environmental impact. Coal, for example, though cheap and abundant, has made cities like Beijing virtually unlivable; coal soot from an electric plant that serves Phoenix is now hanging over the Grand Canyon. Coal-burning is also a disproportionately profligate contributor to greenhouse gases, and thus to global climate change. No wonder that power companies and governments in both the industrialized and developing worlds feel pressure to abandon coal as an energy source when other fuels are available.

But coal is not *inherently* dirtier than other fuels. Rather, the predominant coal-burning process, known as "pulverized coal-fired steam generation," is a fifty-year-old technology with only 40-percent efficiency. Most of the carbon and sulphur inside a lump of coal is simply emitted into the atmosphere as a gaseous by-product of heating up steam to drive a turbine. If a closed-cycle coal-burning system could be developed that pulverized the coal, trapped the carbon as gaseous fuel, and used it in that form, then coal could once again become a major energy source, and legitimate as an environmentally benign fuel.

There's at least one such system under development now, called the integrated gasification combined cycles (IGCC); it may be the harbinger of a new wave of carbon-sequestration technologies that capture the carbon from industrial processes and either trap and store it or make productive use of it. Such technologies would be used primarily in aluminum plants, steel mills, cement factories, and oil and gas refineries—the most egregious emitters of CO_2 gases today. There are also larger-scale proposals to draw CO_2 out of the atmosphere, either through technological means (perhaps burying the carbon in the depleted underground oil and gas reservoirs from which much of it was originally extracted) or by increasing the quantity of marine plankton and forest cover (which naturally convert CO_2 to oxygen). If enough of those methods turn out to be effective and inexpensive, then we could end up using hydrocarbons as fuel for a very long time to come. Conversely, if carbon sequestration proves to be difficult or expensive, then there will inevitably be pressure to move away from coal, oil, *and* natural gas. That adds pressure for other fuel sources to develop—including, very conceivably, a renaissance of nuclear power.

Which brings us to an energy source where I personally stand at odds with virtually everyone I talk to. Currently, there are almost no nuclear plants under construction, even in countries like China, where the demand for energy is growing so fast. Existing plants are seen as bureaucratic, expensive to operate, and targets for terrorism. Most importantly, an accident at the plant (or in handling the fuel) can have devastating consequences for an entire region. Nonetheless, I think it is almost inevitable that nuclear power will come back as a viable energy source, primarily for environmental reasons. This is one energy source that does *not* release greenhouse gases into the atmosphere.

Can all of the objections to nuclear power be met? I think so. The plants are far more resistant to terrorism than most people think. The Israeli air force demonstrated this in 1981 with its destruction of the Osiraq nuclear reactor near Baghdad in Iraq. Iranian fighter jets had already tried to destroy it from outside, with only minor damage. It took undercover work, including setting bombs from within, to destroy it. Other problems, such as the storage of nuclear waste, are gradually being solved; nuclear technology, like other energy technologies, is getting cheaper, safer, and smaller.

There are new technologies, such as "pebble-bed" design, which encases the nuclear fuel in graphite "pebbles," about the size of tennis balls, cools the reactor with helium gas, and operates at higher temperatures.[69]

One can imagine specialized nuclear plants developed in conjunction with other new technologies: for example, as energy sources for desalination plants in China or India, where fresh water is at a premium. We may even see commercial-grade nuclear plants that skip the steam cycle entirely; instead of generating heat to run turbines, they would produce electricity directly from the radiation created by the plant, like a giant solar cell feeding not on light from the sun, but on electron flow from the nuclear reaction.

I'm less optimistic about magnetic fusion, cold fusion, and other fusion power technologies. The physics underlying the process is still poorly understood. Imagine the task of creating a miniature sun on Earth, and then stuffing it into a magnetic beaker, and you may appreciate the challenge involved.

Currently, it's hard to see this green-friendly innovative wave from the inside. For the next several years environmentalists will feel as if they're losing a race—against population, affluence, and perhaps technology as well. In the United States, for example, sport utility vehicles (SUVs) will continue to sell well, despite the fact that they are highly inefficient, fundamentally unsafe gas hogs. It's not just that they allow mothers to carry large numbers of children around suburban subdivisions. They make people *feel* safe and powerful. No green movement will dissuade people from buying them; but all of the technologies and practices we've described will probably go a long way toward reducing the damage they cause.

Indeed, by 2010 or so—maybe sooner—most SUVs will have less impact on the environment than a typical sedan has now. By that time it will have become clear that the tide has shifted. We will be passing out of the era of fossil fuel pollution, into an era when affluence and energy production carry little environmental cost.

Government's Green Priorities

This set of trends is inevitable; governments cannot slow them down. But they can speed them up. By promoting the right kinds of infrastructure change, a few government agencies in any country can do a great deal to move the transition date into a post-fossil-fuel economy forward—from 2030, perhaps, to 2010 or even sooner.

For example, governments can do a great deal to speed up the turnover of capital stock. Although government subsidies tend to backfire, other types of government incentive seem to work more effectively: tax breaks and low-interest loans. Buy a car powered by fuel cells, for instance, and pay no sales tax on it. Or get a $1,500 write-off on your next income tax bill. Or pay no registration fee for the first five years. Or borrow $12,000 at a relatively low interest rate, with the government supporting the loan in the same way that it supports student loans today. The rationale is the same: subsidizing some positive private activities can produce a public good.

This has happened very successfully with refrigerators. In the mid-1970s the U.S. federal government established tax deductions for refrigerators that met energy-efficiency standards. In response the industry created its own "golden carrot award" (an idea suggested by David Goldstein of the California Energy Commission)—a $20 million prize for the most energy-efficient refrigerator. The "Golden Carrot" award, as it's called, has been a major factor in reducing residential energy use, by transforming refrigerators from one of the most energy-profligate appliances to one of the most energy frugal.

Incentives at the production level—particularly tax incentives to make factories, refineries, agriculture, and office buildings more efficient—have also produced good results. There are relatively few decision-makers to persuade to change their habits. We'll probably see most progressive governments moving to set up new and better-designed incentives for encouraging the faster retirement of old factories and plants, and the greening of the technological infrastructure of society. The richer a country becomes, the more likely its politicians will be to do this, and the more interested its industrialists will be in taking advantage of it. One can imagine a time, say in fifty years, when the industrial landscape has become so green that there is little or no loss of property value for homes located on the next block from power plants or factories.

Some resources simply cannot be allocated effectively without government regulation. Water, for instance, will tend to be wasted unless government steps in. We are starting to see this happen in the western United States, where 85 percent of the water supply goes to agriculture. We grow high-hydration crops like cotton, rice, and alfalfa in California—a gross squandering of water to create, in effect, swampland in a desert, paid for by federal farm subsidies. If California's farmers paid what I pay for water, they would use about one sixth of the water they use today. If they emulated the techniques of farmers in, say, Israel, they could grow the same crops they grow today with a fraction of the water.

Water shortages, in short, are caused by government policy; they can be solved more easily through policy shifts than through technology. The same is true of traffic congestion and other infrastructure crises—such as the challenges described in Chapter 4.

Here will be another factor separating "orderly" from "chaotic" nations. In the orderly countries the environment will grow steadily better. There will be two categories of orderly countries: those that are already green (the Netherlands, Sweden, Canada), and those that are becoming green (China, India, Turkey, and the former Iron Curtain countries). In China, environmental quality is at the center of the Communist party's thinking on economic development. They know that it's a prerequisite for sustaining a wealthier society. Residents of many Chinese cities are already coping with significant levels of industry-related disease in their children, caused by coal-burning and other toxic practices. No government, not even China's, can expect to sustain a society that poisons its citizens in the midst of economic growth.

Hence, Beijing is moving rapidly to replace its coal energy use with natural gas, piped in from Kazakhstan, Tajikistan, and Siberia.

Coal is cheaper, but they can't afford its human cost. If they succeed, that will represent a true test of industrial society: Can a billion Chinese people (and another billion on the Indian subcontinent) become genuinely wealthy without poisoning the environment?

The Crisis We Expect: Global Climate Change

There is at least one great environmental crisis that technology can't solve. Nor can government policy—not at this date. Nor can it be prevented. It's not a surprise, because most of us (by now) have seen it coming. The only surprise is the speed of its impact.

The last twenty years in particular have seen an increase in storms, strange cold or warm spells, floods, droughts, and other weather anomalies. They have also seen a growing consensus that the Earth's climate is poised to shift. The accumulation of human-produced "greenhouse" gases in the atmosphere, trapping more of the sun's heat, along with the poorly understood long-term dynamics of the climate, are generally agreed to be the triggering cause. While the timing of climate change cannot be predicted, the prerequisite conditions for it are already here. The National Research Council's report on abrupt climate change, issued in summer 2002, was subtitled "Inevitable Surprises," precisely to make the point that the indicators are too strong to ignore. Some form of global climate change is both unavoidable and unpredictable.

Although the average global surface temperatures are rising, the phrase *global warming* is a misnomer. One can be complacent about global warming; it might simply mean warmer winters in Russia and Canada. And the direct sensory evidence of the last few years seems ambiguous. ("We've had the coldest winter in years this year—how could global warming be taking place?")

A more appropriate phrase is *global climate change*, and it is all too accurate. When people think of climate change, they expect it to happen slowly and gradually. But our knowledge of real-world climatic shifts contradicts that. The fossil record, preserved in cores of ice and mud extracted from near the polar regions, shows how microscopic life-forms changed on a year-by-year basis during great climatic shifts in the past. The pattern is consistent: hundreds or even thousands of years of steady-state equilibrium. Then an abrupt shift, in as short a time as a decade, can alter temperature and rainfall patterns, and ocean currents.

"Given the complexity of how the global climate works," writes neurobiologist William Calvin (author of *A Brain for All Seasons: Human Evolution and Abrupt Climate Change*), it's hard to predict what first- and second-order consequences of climate change would be for various regions, but here are some starting thoughts.

- Increasing temperatures would put tremendous pressures on the availability of water. Key agricultural crops and industries would struggle to survive, particularly threatening life in developing countries.

- Ocean levels might rise, threatening coastal communities.

- Tidal wave activity could create chronic flooding and damage to transportation infrastructure.

- As we have seen with the West Nile virus in New York and the spread of malaria-carrying mosquitoes to northern climates, warmer temperatures would create ripe conditions for deadly diseases like cholera. If people died as a result of these diseases—even a few highly publicized deaths—we could witness a mass migration from the cities and/or the region to live in smaller, highly isolated communities.

Dr. Robert B. Gagosian, president and director of Woods Hole Oceanographic Institution, spells out the possible consequences this way: "Average winter temperatures could drop by 5°F over much of the United States, and by 10°F in the northeastern United States and in Europe. That's enough to send mountain glaciers advancing down from the Alps. To freeze rivers and harbors and bind North Atlantic shipping lanes in ice. To disrupt the operation of ground and air transportation. To cause energy needs to soar exponentially. To force wholesale changes in agricultural practices and fisheries. To change the way we feed our populations. In short, the world, and the world economy, would be drastically different. . . . These changes could happen within a decade, and they could persist for hundreds of years. You could see the changes in your lifetime, and your grandchildren's grandchildren will still be confronting them."

We can't blame human activity or carbon emissions alone for the problem. The emerging consensus is that two large-scale phenomena are interacting. The first is the end of an interglacial period: a shift in global temperatures, which would occur anyway without reference to fossil fuels. Previous such shifts have been responsible for the warm years around 1000 A.D. (when the Norse settled Greenland and traveled to North America) and the "Little Ice Age" of the 1700s (when the Delaware River in New Jersey was icebound as George Washington crossed it and the canals of the Netherlands were frozen throughout the winter). The second phenomenon is human fossil-fuel consumption, which increases the various forcing functions in the atmosphere and may make it more likely that abrupt climate change will be triggered.

It's important, in writing about this potentially serious situation, to distinguish uncertain from predetermined elements. Not one thing in the two paragraphs above is predetermined to take place. Dr. Gagosian's scenario, for example, is based on the idea that the "great conveyor" of deep ocean currents in the Atlantic Ocean might fail to come as far North as it does today. The warm air and water of the Gulf Stream would no longer warm Europe. That theory is very plausible, and could happen even in the next decade. Paradoxically, a warmer climate would drive the shift; as the polar ice caps melted, cold freshwater would pour into the Atlantic and force the currents into new patterns. This has happened before—most recently about ten thousand years ago, in a period called the Younger Dryas, the era of the woolly mammoth. It is far from certain that it will happen now, but there are enough indicators—including a perceptible recent rise in the freshwater content of the North Atlantic—to generate serious concern.

"We are walking toward the edge of a cliff—*blindfolded*," says Dr. Gargosian. "Our ability to understand the potential for future abrupt changes in climate is limited by our lack of understanding of the processes that control them."

What, then, is certain? Only that *something* significant is going to happen within our lifetime. Global climate change is not a problem that future generations might cope with. The climate could abruptly shift, within decades, to some new steady state. Or, worse still, it could oscillate, swinging from warm to cold and back again before finding its new resting place. The new climate could be significantly different from the old—perhaps hotter in some places, perhaps colder in others, and most likely with a dangerous effect on existing populated areas. Even if the various forces affecting climate more or less cancel each other out, and nothing much "happens," there is still the uncertainty. That in itself is a driving force; it means that humanity will not permit itself to stand still and wait.

Global climate change also exacerbates or affects many of the other inevitable surprises in this book. Human migration will dramatically increase as people become refugees from harsh winters, storms, floodplains, or sudden deserts. Necessity will further accelerate technological research and development, especially in transportation, agriculture, habitat, and energy production. (A nuclear power plant won't seem nearly so dangerous if it is the fastest way to produce electricity for heat in a newly frigid city.) Life span and the economy will both be affected; the "hot spots" of Mexico, Saudi Arabia, the Caspian Sea, and Indonesia will come under renewed pressure.

Plague and Denial

Daunting though global climate change may be, there is one silver lining: It is being anticipated. Governments and businesses around the world are beginning to prepare for it—haltingly, but undeniably.

There is another surprise coming, equally difficult to deal with, and probably worse—because no one is prepared for it. We can see the early signs, and we know it is inevitable, but we don't know how bad it's going to be. We are facing the inevitability of another global plague.

There have been two serious global plagues in the last hundred years, which provide some sense of what the next one will be like. The first was the outbreak of influenza in 1918, which killed between 20 and 50 million people worldwide. The second was acquired immune deficiency syndrome (AIDS), which is on its way to killing more than 100 million people. The history of these two diseases gives us a sense of the conditions required for another devastating plague.

- **An incubation point:** a place where the disease can evolve and develop into its virulent form. AIDS, for example, came into the world apparently from a relatively isolated part of Africa, which suddenly came into contact with the rest of the world in the 1970s.

- **A long gestation period** between the time when an individual is infected and the time symptoms appear. This gives the disease an opportunity to spread; human beings, unaware that they are sick, continue their lives as usual and take no precautions against infecting others. This was the pattern with influenza, which we now know (through recent genetic research on tissue taken from people who died in 1918) was contracted in some victims as early as 1902, sixteen years before the symptoms appeared.[70] It is also the pattern with AIDS. Ebola, which is a truly horrible disease, is mercifully swift to appear in its hosts; they have little opportunity to infect others, and it tends to run its epidemic course quickly.

- **A large, uninfected, and infectable population—without immunity to the disease.** The more unfamiliar the disease—the farther it travels from its original site—the more likely it is to encounter people who have not built up an immunity to it. That's why the next plague is likely to be either an entirely new strain of virus, or one that disappeared so long ago (like influenza) that an entirely new human population has emerged with no genetic immunity remaining.

- **A distribution system that brings the disease to the infected populations.** One reason for the return of epidemics, despite massive improvements in public health since the turn of the century, is the availability of inexpensive air travel. Airplanes carry people, pets, and microbes to parts of the globe where they have never been before. In fact, the person who unwittingly brought AIDS to the West was an Air Canada flight attendant, a gay man who later became known as "Patient Zero."[71]

- **A consistently reliable form of transmission from one person to another.** AIDS continues to spread because people exchange fluids—either through transfusions, needle sharing, or sexual activity. Cut out the forms of exchange, and the disease stops spreading. Influenza was much easier to spread; it merely required breathing in the same room with someone else who had it.

- **Ignorance about the disease.** AIDS was originally known as the "gay cancer," and treated as such. Influenza was assumed to be caused by a bacteria, and a vaccine was created accordingly. But, in fact, it was a virus. This ignorance was one of the factors that prolonged ineffective treatment and allowed the disease to spread.

- **Denial of its seriousness.** Public Health Commissioner Royal Copeland of New York City issued this statement in 1918: "The city is in no danger of an epidemic. No need for our people to worry." In his book *And The Band Played On*, AIDS historian Randy Shiltz documented similar statements from both government and gay community leaders in the early 1980s. More recently, we've seen that kind of self-deception from politicians in Russia, China, India, Southeast Asia, and South Africa. The longer people refuse to recognize the plague as a public health crisis, probably requiring intensive research, new forms of quarantine, and changed behavior, the more time it has to spread.

All of these conditions exist today. The chances of some new disease evolving to meet these conditions are so high that it is practically inevitable.

AIDS itself, as noted in Chapter 5, will inevitably spread further. Currently, 100 million people throughout the world are projected to die of AIDS during the next thirty years. AIDS is difficult enough to control; each new generation of young people develops a cohort of experimenters who imagine that they can get away with unsafe sex. But the disease that will strike us will probably be communicated more easily than AIDS; it won't require direct blood exchange or sexual contact. Imagine, however, if a disease like AIDS mutated into an air- or waterborne form, or a form carried by insects (like malaria). Imagine if it had a long gestation period, so that people could carry it unknowingly from one country to another on airplanes. Hundreds of millions of people could be infected by such a disease in a matter of weeks. We don't know for sure if it will be as devastating as that; and we don't know for sure exactly when it will come. But we do know it is coming.

It could be an entirely new disease, like Ebola or AIDS. Or it could be an old disease, once thought eradicated, that has evolved into a new drug-resistant form. We're already starting to see antibiotic-resistant strains of staph and tuberculosis; there is a form of malaria returning that is resistant to quinine. Or it could well be a new variety of influenza; no one has quite figured out why the influenza epidemic stopped spreading, or why it hasn't come back. It might also be generated by an accidental overreach through genetic engineering.* And we cannot dismiss the possibility that terrorists will generate and spread the next plague.

Whatever its source turns out to be, it will be immensely destructive. Probably the most tragic effect of both influenza and AIDS has been their impact on children. Influenza left hundreds of thousands of orphans. There are currently 14 million AIDS orphans in Africa alone. Many of them have watched their parents die. Many have no one left to care for them. In Botswana one third of the educated adult population is infected and will die of AIDS in the next ten years. The country consists of very old and very young people, with relatively few others surviving. Now imagine if a disease had the same effect on the United States and Europe.

There will also be serious economic effects. We saw a hint of this a couple of years ago when an Air India flight was turned away from Britain because of a cholera outbreak in Calcutta. Imagine the impact on world trade and on education if flights had to be turned back around the world; if countries had to be quarantined.

But will the new plague be like influenza, remembered for generations as a one-time scourge? Or will it be a threshold marker, like AIDS, after which society is never the same? That depends on how well we are prepared for it.

Currently, we are extremely poorly prepared. Public health systems in most parts of the world are already overtaxed. The data systems aren't in place to track and compare epidemiological information. Water and sanitation systems around the world, while improving, are reprehensible in many nations; electrical and trans-

*The evidence so far suggests that genetic engineering, even with food, will not lead either to an ecological or a health crisis. A number of mistakes have been made, and genes have been artificially released—with minimal consequences in the end. To be sure, there are risks; but so far the law numbers has prevailed: potential accidents have been prevented simply by virtue of the altered genes being massively outnumbered in a very large pool of experiments.

portation infrastructures are far short of what they would need to be to combat the disease.

There are two things that need to be done—probably in a partnership among government, business, and foundations. The first is to improve public health facilities, particularly for sanitation and water. The second is to invest in detection systems. There needs to be a much better network available for hospitals to share information; and educated people need to be subsidized to continually make sense of the information and look for cues that a new disease has emerged—while its population base is still relatively small.

Calamities tend to reinforce each other. If we are unprepared for floods or chilling weather, then that will make people all the more vulnerable to new infectious diseases. Conversely, if we have the infrastructure in place to help people prevent the spread of infectious diseases, that will also help us meet the challenges of global climate change. Investing in these kinds of infrastructure now might seem to risk putting a drain on the economy—but in the end it might turn out to be the single piece of foresight that saves the economy from collapse in a global health crisis.

The Final Calamity

There is one more inevitable surprise that must be mentioned—though it occurs on a scale so large that it hardly fits with anything in the rest of this book. Sooner or later some civilized part of the Earth will be hit (or at least threatened) by an asteroid.

We don't know when that will happen. It might not happen during our lifetimes. It could take place five years from now, five hundred years from now, or five thousand years from now. We might have years of warning; or we might have no warning at all. The Earth just had a close call with an asteroid that came at us from the direction of the Sun, and was literally undetectable until just before it passed by.

This isn't precisely an "environmental" issue. Nor is it something that can be prepared for (though, arguably, NASA and every aerospace company should be doing precisely that). There are plausible technological means that could deflect an asteroid or comet. But if it's large enough, and it hits the wrong part of the Earth, it could virtually destroy human civilization. We at least ought to be watching carefully for them.

The last such event took place in 1906. A fragment of a comet entered the atmosphere and blew up a few miles above a remote part of Siberia called Tunguska.[72] One town was seriously damaged; one forest was flattened. Fate placed the asteroid over one of the most remote locales on the land surface of the Earth. Had it fallen instead over a major city—New York, London, Mexico City, Beijing, Tokyo, Delhi—the metropolitan area and most of its inhabitants would have been flattened.

When events like that take place, how do you look at them? Do you see them as the hand of God? As isolated events? Or as a warning of something that may take place again?

CHAPTER 9

Inevitable Strategies

In 1993, the University of San Diego mathematician and science fiction writer Vernor Vinge proposed that humanity was entering a no-return transition point that he called the "singularity"[73]: a point after which human experience would change forever. Or, as Vinge put it: "a point where our old models must be discarded and a new reality rules." Vinge (and others, like the computer scientist Raymond Kurzweil) have posited that the tipping point will come when computers are developed with processing power on a par with human intelligence. From there machines will begin to design and build their own machines, outpace the human ability to understand their purpose and processes, and direct the course of progress ever after. "Any intelligent machine would not be humankind's 'tool,' " wrote Vinge, "any more than humans are the tools of rabbits or robins or chimpanzees."

I don't think the evolution of computers beyond human capability is inevitable. It's certainly plausible, but I don't think it's predetermined to take place sooner than 2030, as Vinge does.

And in a sense it doesn't matter. Whether or not that specific technology is realized, a singularity is approaching anyway. It is the third such singularity in human history, and it will be upon us within the next twenty-five to thirty years.

The first such singularity occurred about eleven thousand years ago, and took several millennia to travel around the Earth. It was a Great Transition for humanity: the advancement of the species from a survival strategy of hunting and gathering, to a state of civilization based on agriculture. That civilization, by the time of Jesus Christ, was characterized by centralized authority (often in the form of a monarch), slaveholding (just about every agricultural civilization had slaves of one form or another), commerce (often in the form of markets), and literacy for the elites.

Beginning with the Europeanization of movable type, and ending with the technologies of the mid-twentieth century, there was another singularity: the industrial revolution. This Great Transition might perhaps be described by a simple listing of inventions: clocks, telescopes, guns, motors, steam engines, telegraphy, railroads, electricity, automobiles, telephony, radio, the submarine, the airplane, the rocket, the television, the computer, and the atom bomb. But there was much more to this Great Transition than technology.

It took several hundred years for the mutual interaction among technology, economics, local politics, geopolitics, media, culture, agriculture, medicine, religion, and patterns of community development to interact and produce the kind of civilization that we think of as "modern": distributed authority (with many large-scale democracies), machine automation, large-scale organizations as commercial entities, relatively widespread education, a shift in attitude about wilderness (no longer a threat to humankind, it was a resource that needed protection), the birth of modern medicine, and ultimately the terrors of fascism and communism. The world in 1925 was a very different place from that of 1850—and even more radically different from that of 1650. But someone looking back from 1925 could see, in effect, how the pattern of civilized change had progressed.

The period since World War II, for all its change, has been a time of relative stability. Alvin Toffler's *Future Shock* suggested that the pace of change was speeding up; but in truth it seemed to level off compared with political and cultural upheavals in the years between, say, 1850 and 1930. Space travel, the personal computer, and the cellular phone are civilization-changing technologies; but they cannot compare with the impact the development of the electric light, the broadcast signal, and the automobile had on the average person. Nor can the breakup of the Soviet Union, for all its significance, compare in impact with the rise of democracy two hundred years earlier.

Each of these great transformations has followed a similar pattern based largely on demographic and scientific-technical factors: how many of us there are, our age distribution, where and how densely we live, what we know, and what we know how to do. These are the human dimensions and capabilities that lead to the economic and political process of development and transformation. These fundamental transformations are often linked to huge ecological changes as well.

In the first "Great Transformation," for example, we saw the warming and stabilization of the Earth's climate accompanied by the birth of agriculture and village life. As greater numbers of people survived and lived longer and more securely, civilization began to develop. During the second Great Transformation, which may have been linked to the beginning and end of the Little Ice Age, we developed a new world view, new technologies, and an increasingly integrated world economy and polity characterized by larger and more complex organizations. Population took off after 1890 with the advance of agricultural science and a warming planet.

Looking ahead to the third Great Transformation we see many more people living far longer and global population growth decelerating to its peak in the mid-twenty-first century. The revolutionary advance of science and technology, especially in the life sciences, will create new, fundamental, and controversial possibilities for our species. The economic potential to lift billions out of poverty will be within our grasp. And a new kind of political order will result. Yet, as in the past, we are increasingly likely to find ourselves challenged by climate change—this time of a very abrupt nature. During the next twenty years, there may well be as many disruptive and overwhelming changes for civilization as there were in the first two Great Transformations.

Why would this happen? Because of the magnitude of changes in a variety of arenas:

- A dramatically increased life span, and its effect on human identity, capability, and community;

- New patterns of human migration, which either fragment or unite humanity in new ways;

- The return of a reliable "Long Boom" with global investment, expanded productivity, and the unprecedented opportunity for people around the world;

- A dominant global military and economic superpower—the United States, with unchecked reach and a potential for political capriciousness—such as has not been seen since the Roman Empire;

- A consortium of nations bound together not just by their intentions but by their common need for lawful collaboration;

- A set of disorderly nations with the capacity to unleash terror, disease, and disruption on the rest of the world;

- Technological capabilities that include new materials and machines, which could expand the power of computers by orders of magnitude, and permit people to reprogram reality;
- Pollution-free and inexpensive energy sources that release humanity from its addiction to fossil fuels;
- A restoration of nature under human stewardship, alongside extreme and unavoidable danger from a new plague and global climate crises.

Any of these in itself would be a significant change. But the greater significance comes from the ways in which all these changes will interact together.

Anticipating the Cumulative Effects

Imagine reading a book like this in, say, the year 1895. It includes details on the nascent science of electromagnetic transmission, and informs you that an Italian named Guglielmo Marconi has just transmitted a form of telegraphic code through the air. Someday, it says, we may do the same with sounds and pictures.

Then it lists other inevitable surprises in the works: electric lights, motion pictures, automobiles, and even the airplane. Less than fifteen years from now the Wright brothers will be launching their airplane-constructing business. Perhaps the book might even describe the inevitable results of new infrastructures under construction now: subways, sewage systems, and water supply systems, as well as the even newer infrastructures needed for electric power transmission, telephony, and the automobile. Perhaps it would even be prescient enough to recognize those technologies that will turn out to be insignificant (the Zeppelin) or short lived (the telegraph and the steam engine).

And the book would not stop with technology. It would contain the result of conversations with political leaders, military leaders, financial leaders like J. P. Morgan, and perhaps social pioneers like W.E.B. Du Bois and Susan B. Anthony. It would explore the growing phenomenon of international trade, the opening of a canal across the Isthmus of Panama, and the exploration of the North and South polar regions. It would describe advances being made in cure for malaria, and the changes in education and longevity for wealthy people in London, Paris, and New York. It would talk about the radical concept of world government, and foreshadow how all nations could come together in a global league—not in a reaction to war (for that wouldn't be foreseen yet), but perhaps because of world trade and the well-established peace.

There were, in fact, a number of such books. (One of the most famous was titled *Looking Backward*, by Edward Bellamy.) They were so popular that a publishing genre was named to describe them: Utopian fiction. They varied quite a bit in their prescience and approach. But even the best of them missed a great deal of the future.

It was difficult to foresee, for instance, how the automobile would affect disease control by reducing the number of animals in cities. No one could have identified the way in which modern bookkeeping techniques, the telephone, and the airplane would empower the rise of multinational corporations; or the impact of industrialization on climate change, or the relationship between the computer and the science of genetics, or the end of European colonies and the surge in population growth in what was then called the "colonial" world. Few could have imagined the ways in which the peace of the Edwardian era helped create a cultural readiness for war, and the ways in which the brutal experiences of World War I ensured that the peace that followed would be incomplete.

In short, no matter how skilled a book of 1890 might have been at extrapolating and foreseeing individual inevitable surprise, it would have been almost impossible to delineate the most important inevitable surprise of all: the second-order effects that naturally occur as these changes reinforce and affect each other in dynamic, cumulative, self-reinforcing ways. So they were not just individual developments, but fundamental shifts that made the world of 1930 radically different from the world of 1900.

What, then, are the cumulative effects of the inevitable surprises that have been described in this book? What are the ways in which these "first-order effects" will influence and reinforce each other? What are the inevitable second-order effects?

First, the more extreme the first-order effects become, the more explosive the second-order effects will be. That is the reason, in fact, that the transition of the next few years will be as significant and world changing as the transition from 1650 to 1950.

Second, the next transition will take place over a much shorter time than the last one—perhaps over the course of thirty years instead of three hundred. If this is the case, by the end of the lifetime of most people reading this book, it will have happened.

Third, the stability we feel today is rapidly going to disappear. Utopian authors at work in 1890 were writing in the midst of the "Gilded Age": a time of great stability, when it seemed peace and prosperity could go on forever. Looking at the changes ahead of them, they might have written, "The world seems stable now, but it will be a surprise if that stability continues." And they would have been right. The next sixty years would bring two world wars, a massive depression, an end to European colonial empires, and many other changes.

The same is true of the next sixty years ahead of us today. The magnitude of change will be so large and disruptive that, if there should be economic and political stability in the world, it will be a great surprise. The potential for progress is enormous, but the potential for disruption is equally great.

The World of Maximum Surprise

Let's consider what it will be like to live in the year 2030 if nearly all the surprises in this book come to pass before then.

As you walk down a typical street in a city of 2030 you will see far more older people, at much more advanced ages, than you would ever have seen in 2003. You might see men and women of seventy doing things that formerly only people in their twenties and thirties did, such as holding hands in public, or pushing baby carriages containing their own children. Some of these people might have a kind of synthetic strength, youthfulness, or other physical attributes that would have looked distinctly peculiar in 2003, but which are now taken for granted. This is because of the more radical and interactive impacts of the new biomedicine. Will this lead to a large number of significantly modified human beings? If so, they would probably include some people who have undergone genetic treatment as adults, and others, younger people, whose genetic code was modified while they were still embryos. They may be smarter, bigger, stronger, longer lived, and more disease resistant than their parents. Some of them may be bioenhanced warriors, fighting against terrorists and shock troops from the disorderly parts of the world.

With each passing year the aged will be getting younger, so that the centenarian of 2030 (including some readers of this book) will resemble a sixty-year-old of 2003. If you are one of these people, you will find a variety of genetic therapies available for you; diseases like Alzheimer's, diabetes, heart disease, arthritis, and many more will claim fewer and fewer victims.

Will only a wealthy few benefit? Or will the cost continue to fall, making the new technologies and remedies available to an ever-increasing number of those who need it? The answer may well vary in different parts of the world. If Europe can hold off the demographic tensions that threaten its unity, and preserve its social welfare ethic, then it may come to finance that ethic, in part, by recreating itself as a center of new therapies. Doctors and patients alike would be drawn there by not just the high quality of life, but by the secular nature of its society; religious opposition to many of the therapies could make Europe the center for the new medicine.

This is an especially plausible prospect if the migrants to Europe, from the Middle East, India, and Russia in particular, include people with the will and talent to work in this field. Certainly the number of aging people there would provide some impetus; by 2030 the number of centenarians will be growing fast especially in Europe, the U.S., and Japan. And they will be younger and more vigorous than the few who feebly stumbled over the century finish line in the past. All of this might help energize the European economy, particularly if Europe's proactive leaders, emboldened by their success with European union, recognize the potential benefit of investing in their population. They might see high-tech biotechnology as one arena where private enterprise (starting with the Swiss biotech industry of 2003) and public government (starting with the European Union bureaucracy) can work together to manage the necessary institutions effectively and still promote research and development. We might then see a high-tech bioentrepreneurial revoluton in Europe.

At the same time, we might see whole European cities evolve into ghettoes for Muslim and African immigrants, virtually walled off from the rest of the continent, and festering with crime, disease, and random violence like the American ghettos of the 1970s and '80s. If that happens, then the ship of European integration might founder on this rock of immigrant-related tension. Unquestionably, the faces on the street in any European city will be increasingly more multiracial. But will they be prosperous and content or impoverished and angry? The answer is not certain. We know that the typical city in the United States will be similarly diverse, but we also know that it will not have nearly the same levels of ethnic tension. We'll already be accustomed to seeing the American way of life as a culture with widespread Hispanic and Asian components, so much so that we won't always recognize them as such. Fajitas and lo mein will be long-standing American food staples in 2030, just as bagels and pasta were in 2003.

In the "disorderly world" the falling birthrates will lead to an ever higher proportion of elderly there as well. Foreign aid to developing countries will have shifted toward new problems, including the needs of a generation of AIDS orphans now entering their thirties and forties. Will there be any feasible way for countries that have endured thirty years or more of disorder to climb out of that hole? If not, one can imagine the future that many people fear coming to pass: islands of prosperity in a sea of poverty and despair. Gated communities in the rich world will hold back the tides of crime and misery, while more and more countries lose ground.

Or perhaps the gathering wave of prosperity that exists in the wealthier parts of the world will be enough to lift up all of humanity. If the current growth rates are sustained, then the world will be about one and a half times richer than it is today. The three big drivers of wealth—new productivity, new globalization, and new infrastructure—are all likely to increase, with varying rates of development. In the traditional dynamics of creative destruction we will transform old industries and create entirely new ones. The new older and more experienced workforce will be enabled, with ever more powerful technologies, to be increasingly more productive. If there is a political will (or a philanthropic way) to ensure that this wealth is distributed more broadly, then we might see an ever-widening circle of prosperity. More and more poor countries could begin to develop a significant middle class, while the rich countries would make great progress in reducing poverty within their boundaries. It won't happen on its own, but we know that the necessary ingredients for such a future, including the political will, will exist.

Prosperity will depend, in part, on bioelectronics and bioindustrial processes that will spark a new computing and industrial revolution. Imagine building things the way that nature does. And of course, these new life-science-based technologies will contribute to making the earth greener and slowing the use of nonrenewable resources. If the new scientific revolution is as broad and deep as I believe it will be, then it will also include far more powerful computers, based on "quantum information sciences" (as we'll call it). This, too, will be a much more global enterprise—based in part in American research universities, but with outposts and researchers all over the world, especially in Russia, China, and India. Any accomplishments in this realm will accelerate accomplishments in all other fields—for instance, it would enable us to manage the complexity of intelligent highways, with computers that map and control the position of every vehicle in real time. Or it would enable biologists to model the unimaginably complex folding of protein molecules, leading to still faster advances in medicine. Ultimately, it might signal the arrival of artificial intelligence.

That would be enough to propel us into the "days of miracles and wonders," as Paul Simon put it. But science and technology could take us farther still. If a new synthesis emerges in physics, we will begin to see new technology that operates on new physical principles. This could include new ways of generating energy, new modes of propulsion and travel (including space travel), new methods of computing and storing information, new modes of communication, and new kinds of sensors. Perhaps the future of energy does not lie with any of today's technologies and fuels but with entirely new domains of science such as dark energy. Perhaps the future of aviation lies not with the aerodynamics of lift, but rather with the physics of antigravity. We may think of these as fantastic ideas now, but if any of them come to pass, it will dramatically accelerate the shift in all aspects of society, if only because each of these new technologies will require new infrastructure, new financing, new cultural adaptation, and new corporate and government sponsorship. If China, for example, becomes the center of research in the new physics, that could mean the first serious challenge to the American political hegemony of this new era.

Geopolitics may well be the hardest domain to envision for 2030. It is virtually certain that the United States will still be the single strongest nation on earth in both economic and military terms. But will it be a more peaceful or a more troubled world? Perhaps the Bush administration is right. What the world needs is "tough love." The U.S. thus becomes a reluctant but conscientious sheriff, shielding the orderly nations from destructive forces emanating from the chaotic and failed parts of the world. In such a future the U.S. would naturally become more adept in cultivating the support developing and imprimatur of the other nations, especially the great powers and major international institutions. As that happens the other nations, even some of those currently opposed, might come gradually to support the U.S. role as the peacekeeper and guarantor of the new order.

Such a benign, stable, and peaceful outcome relies on two key assumptions: that the problems are actually addressable and the U.S. is competent in the execution of its aims. A wealthier world may not be sufficient to overcome the passions inflamed by a battle of theologies. We may by then be enduring the twenty-fifth year of a new Thirty Years' War, exacerbated by the spread of biotechnology that could arm new forms of terrorism. Indeed, though the U.S. military has had mostly a solid string of success since Vietnam, that is not predetermined to continue—especially because we are confronting new and uncomfortable terrain in the war on terror. Success is not guaranteed and a protracted and widening conflict is not implausible.

Even if this war between Christianity and Islam does not materialize, there are still plenty of hot spots to create trouble and many competing interests to create tensions. It is not hard to image a turbulent world of 2030 in which the U.S. is powerful but isolated. Now in its thirtieth year of "rogue superpowerdom," America could face skilled coalitions of denial (perhaps *still* led by French and German leaders, perhaps by other Europeans or non-Europeans) who have learned to use their soft power and the international rule of law to block U.S. interests. In such a world there will often be hot spots needing attention. It is in these poor benighted countries that the U.S. will play out its conflicts with the other great nations. In this sense it will resemble the Cold War surrogate struggles between the Soviet Union and the United States in places like Africa, Southeast Asia, and Latin America. If the world of 2030 is dominated by chaos and conflict, then poverty and violence will feed on each other in a downward spiral of self-destruction.

Here lies the challenge, and the choice, of the world ahead. Will the world be more like China, in which a one-party state creates stability and prosperity that go hand in hand? Or like India, in which a disorderly democracy has produced a mixed record of economic progress? Or will other models for effective political economy evolve? We still don't quite know how to generate economic stability in a world of turbulence and inevitable surprises—but there is some reason to think we may be able to pull that rabbit out of the hat after all.

If we are successful in bringing more people out of poverty in a world of great instability, change, and tension, then prosperity may begin to feed on itself and accelerate. The resulting 2030 would be a cleaner, greener, richer, safer, and less divided world. The compounding second order effect producing enormous progress would certainly surprise all of those who have a much darker vision of the future. In the end it comes back to our ability to prepare for and learn from the surprises described in this book. They may be inevitable, but our responses to them are not—and it is our responses that will make all the difference in the quality of life, or lack of it, in the world of 2003.

Preparing for the Future

The next logical question is the same that a reasonable reader of a book about the future might have wanted to ask those imaginary authors of 1890:

As someone who cares about my business, my locale, my community, my country, or my family, what are the precautions I would want to have in place?

In other words, you can't predict the chaos and turbulence to come. But how can you best prepare? What foresight can you cultivate, so that when this level of instability comes, you, and the people you care about, are ready for it? How can we learn from the last Great Transition to be better prepared for this new one?

I have learned that several answers are available from years of helping organizations anticipate the future in scenario practice.

- Build and maintain your sensory and intelligence systems. That doesn't just mean technological systems. It means the continued kinds of "strategic conversations" in which you and your cohorts and colleagues keep looking around to observe and interpret the interaction of forces that might affect you, your enterprises, and your communities.

This seems obvious, but it's surprising how many politicians, educators, and businesspeople I have met who do not make time for it. Over the years their ability to observe and interpret the world around them atrophies. In a singularity like the one approaching us, fine-grained awareness of the world outside your own organizational boundaries will be a paramount aid to survival.

■ Cultivate a sense of timing. When you see an event approaching, make a point of asking: How rapidly is it approaching? When could it occur? How far in the future?

■ Identify in advance the kinds of "early-warning indicators" that would signal that a change is rapidly upon you. For instance, if you are a foreign investor, what are the early signals of potential financial crises? You know they will occur in China and India—what do you look for there? If you are a technologist, what kind of funding will be evident in your arena first, before it attracts financing from elsewhere? If you are concerned about climate change, what represents the next big warning sign? And how do you distinguish it from run-of-the-mill climate variation?

Once you've identified these signals, keep an eye out for them and be prepared to act when you observe them. This is one place where my colleagues and I use short-term scenario exercises: "If we saw such a signal, what could it mean? And what would we do in response to it?" In 1997, when the financial crisis hit Southeast Asia, the U.S. Treasury had already undergone the kind of firewall-building exercise necessary in Mexico in 1994—which made it possible to move rapidly to contain the crisis, so that it did not ripple into China, Korea, and Japan.

■ Put in place mechanisms to engender creative destruction. The institutions, companies, agencies, political parties, and values of the past may turn out to be moribund and counterproductive in a new historical environment. Are you prepared to discard them? More importantly, have you practiced discarding them? What processes, practices, and organizations have you actually dismantled in the last year or two? If the answer is none, perhaps it's time to get some practice in *before* urgency strikes.

Creative destruction is not simply a matter of getting rid of old baggage. It means learning how to mitigate the costs. There is inevitably a fair amount of disruption to communities, the abandonment of secure livelihoods, and the severing of deep relationships. You cannot keep those old institutions for the sake of convenience; you need the creativity that comes from releasing them. But unless you can ease the pain of disruption, you will engender fierce resistance. Moreover, the pain of disruption tends to fall disproportionately on the "20-to-40-percent group": the hidden population of lower-level employees on whom the revival of the economy depends (as we saw in Chapter 4). Unless you can help them bear the consequences of disruption, you may cripple your ability to recover.

Note how many of the most successful businesspeople and politicians of the past twenty years have been successful at this, including the last two American presidents (both of whom arguably rode to success by largely discarding the previous identities of their political parties).

■ Try to avoid denial. When an "inevitable surprise" comes along that makes life difficult for you or your organization, do not pretend that it isn't happening. This book is full of examples where leaders exacerbated a problem by trying to deny its significance: AIDS in Africa and Russia, the telecommunications "last mile" problem, and the potential severity of global climate change.

Unfortunately, most standard corporate or government planning is a recipe for denial. The standard operating procedure is to talk about the various futures that might lie ahead, pick the one that seems most likely to happen, plot the course accordingly, and maybe build in a few exigencies. Having done this, the planners (being, after all, human beings) are naturally prone to discount any signals from the outside world that contradict the outcome they expected. The very fact that a future feels "likely" should make us skeptical of it. Chances are, we are drawn by our own limited worldview and predisposition to assume that what we expect to happen, will.

By contrast, when a future feels particularly wrong or discomfiting, and your first impulse is to say, "That would hurt us if it happened, but it won't happen," that's a signal to pay closer attention to it. Something about that future is trying to break through your mental blinders, and if you deny it or ignore it, you may well inadvertently help to bring it to pass. Indeed that kind of denial may have caused the NASA leadership to deny the potential for catastrophic failure despite contrary evidence on the *Columbia*.

■ Think like a commodity company. Most goods, services, and financial sums that are traded are commodities—no one has a monopoly on producing them, and therefore they are subject to swings in supply and demand at any time. This includes not just real commodities, like oil, gold, and wheat, but also stock prices, tax receipts, and trade revenues. It's all too easy to believe, on a price upswing, that this time is different, and the commodity will rise forever. But sooner or later the price will hit a peak. Those peaks can come suddenly, and the aftermath can be a steep and highly disruptive fall.

As a resident of California I'm seeing this take place now with my own state government. During the "dot-com boom," tax revenues poured in, as both companies and individuals grew rich. The governor and the legislature piled on services and spending accordingly. They didn't plan for the day when the revenues would drop, and they must therefore now deal with a $45 billion budget shortfall.

Traditional commodity companies understand this temptation very well. They know that there will be fat years and lean years and the surplus from the growth periods allows you to survive the deficits of the down years.

■ Be aware of the competence of your judgment, and the level of judgment that new situations require; and move deliberately and humbly into new situations that stretch your judgment. Every successful individual and organization has an integrated core of judgment—not just knowledge, but the ability to make wise decisions quickly in a particular field—that lies at the heart of success. When times are turbulent, the temptation to move outside that knowledge to take advantage of outside opportunities is great. Those are the risks that often get you into trouble.

My own education-by-fire in this principle came at Royal Dutch/Shell. In the mid-1980s Shell was flush with cash: $13 billion of it, a legacy of a decade of high oil prices. Along with a brilliant young Shell treasurer named David Welham, I proposed risking some of it in international currency arbitrage in what seemed a very low-risk way of using our short-term funds more efficiently. We could lend it to big banks like Citicorp at a favorable rate, and they could use it for currency arbitrage—for which they needed huge piles of cash. We would make a tenth of a point per day, and they'd make a half a point; on a few billion dollars this would provide an income of hundreds of millions of dollars at virtually no risk.

The decision went up the hierarchy to Bill Thompson, the managing director of Royal Dutch/Shell for finance at the time, who was David's and (indirectly) my boss. Bill vetoed the idea. "We're not a bank," he said. "We've got to manage our money appropriately, but our business is not making money with our money. We are, as management, incompetent to make the necessary judgments."

I was extremely disappointed. But I didn't realize the wisdom of that decision until a number of years later. At that point a failure of control had opened the door to strategies similar to those Bill had rejected. And a Shell currency trader working in Tokyo lost $900 million in one day. It was the biggest currency trading loss of any sort for any company in history. It happened because the controls and judgment were simply not in place as they would have been in an experienced bank.

Bill was right. I was wrong. Because we had figured out our oil strategy so well, I was supremely confident. I thought we could do anything. I'm not saying that I would have lost $900 million, but I am certain that I didn't know enough about playing the currency game to set up the necessary controls. And I've observed, all too many times since, that would-be innovators get their way too often, without someone as smart and wise as Bill Thompson to stop a foolish new idea.

■ Place a very, very high premium on learning. Most failures to adapt are, in effect, failures to learn enough in time about the changing circumstances. And there will be more to learn in the future. If advances in science and technology are any indication, work will be increasingly knowledge intensive, and the value of scientific knowledge in particular will be all the greater.

Unfortunately, most Western societies have approached education ideologically. There has not yet been a genuine consensus, among educators and budget-setting politicians, about how children and adults learn, and about how best to set up schools. Until such a consensus is reached in the most pragmatic, nonideological way, we are unlikely to see a functional education system in most countries. Instead, we will have what we have now: various splinter groups arguing that their favored approach is best for schools, and no solid way to compare the results. (Standardized tests measure only a very small part of the capabilities that people need education to gain.) This is an extremely dysfunctional way to deal with the future.

■ Place a very high premium on environmental and ecological sustainability. This is not just a global political and environmental issue; it is a vehicle for high-quality integration and development. You almost have to run an organization that follows this path to recognize how valuable it is; it focuses attention on the "side effects" of your actions, in ways that are extremely useful.

■ Place a very high premium on financial infrastructure and support. Individuals need safety nets and insurance against crises. Organizations need to build in safeguards and help individuals build the financial infrastructure they need. And society as a whole will need to watch out for the interests of the "20–40-percent" group, for whom no one else typically is.

The risks are greater than we think. In the future, people at all three levels will need safety nets in a way that hasn't been true before. And organizations will need to muster profits and use them wisely. Do you have the kind of portfolio of income and assets that will help you weather the storms to come? Do you have enough profits to fund your transition into the next stage of your evolution, whatever that turns out to be?

- Cultivate connections. In the world of 2025 people will be inevitably in contact far more regularly and comprehensively than they are today. Quantum computing, universal broadband, longer lives, globalization, and clean, green energy will reshape our world toward far greater interconnection. Are you prepared for this? Do you have the kinds of deep, candid connections that will help you ride through the next transition without having to ride alone?

Taking the Long View

At the beginning of this book I used river rafting as an analogy for the future. But perhaps sailing is a better metaphor. One wave after another is going to hit your ship, and you have to be able to react immediately to them. But are you merely stumbling from one wave—one crisis—to the next? Or are you the master of your fate, moving toward a point that is a long-term vision of your own?

The world will not make those decisions for you. The future will almost certainly be a prosperous world, with lots of new technology. But it will not have solved the problems of poverty and "have-and-have-nots;" indeed, those problems may be more extreme than they have ever been before. It will have quantum computers and remarkable new forms of infrastructure, but it will also be struggling against outmoded infrastructure and a capital base that will not inhibit the old infrastructure giving way to the new. It will have broken the barriers of aging and genetic engineering, but it will also have rampant plagues, either naturally developed or spread through terrorism, that could be as bad as any in human history. It may fall to you and your organizations to help solve some of these problems, or others. Or you may simply seek to thrive and keep yourself and family afloat. But you will achieve neither easily unless you choose where you are going.

Consider the difference between a trading company and an investment company. A trading company counts itself successful if it just comes out ahead, day by day, in its accounts. An investment company knows that success depends on being able to reach a significant long-term goal: building cars, opening a market, or creating some kind of new infrastructure (or whatever it may be). Trading companies are often more comfortable to be part of the present moment. But they invariably get wiped out. The investment companies, just by virtue of pursuing a long-term goal, have been building the capabilities, protections, and judgments they need to survive; while those who live by immediate circumstances, die by them as well.

The great risk of our time is being overtaken by inevitable surprises. When we don't have a sense of direction and purpose, we can easily be swept away by events. We have an example in recent history to consider: it's the first half of the twentieth century. Had leaders of the world been willing to think ahead more, they might have avoided two world wars, a depression, millions of deaths, and a half century of global disruption. Arguably, had that happened, we wouldn't be facing the kinds of challenges that daunt us today.

The second half of the twentieth century indicates that we may have learned that the cycle of progress and disruption is not predetermined. It is possible to break it. It is possible to see beyond immediate events, hold fast to long-term directions, and maintain the resources to manage the consequences of disruption. We can't stop disruptions from happening, but we can cope with them far better than we have in the past.

There is no recipe or playbook for doing this. There is only the ongoing knot of life to unravel. Perhaps the string that is the easiest to pull first is the string of inevitable surprises.

NOTES

Chapter 1. Inevitable Surprises

1. Art Kleiner, *The Age of Heretics: Heroes, Outlaws, and the Forerunners of Corporate Change* (New York: Doubleday, 1996).

Chapter 2: A World Integrated with Elders

2. Mark Lane, Donald Ingram, and George Roth, "The Serious Search for Anti-Aging Pill," *Scientific American,* vol. 287, no. 2, August 2002.

3. An interview with Dr. Michael West by Life Extension, "Conquering Cloning with Aging," April 27, 2000, available at *http://www.lef.org/featuredarticles/apr2000_clon_01.html?GO.X=0\&GO.Y=0*

4. Jeff Donn, "Leukemia Drug Restores Some Color To Gray Hair," Associated Press, August 7, 2002.

5. Robert A. Freitas Jr., "The Future of Nanofabrication and Molecular Scale Devices in Nanomedicine," *Studies in Health Technology and Informatics,* 80, July 2002.

6. "Who Is America's Oldest Worker?" PR Newswire, April 18, 2001, Alice Ann Toole, Green Thumb, Inc.

7. Andrew D. Eschtruth and Jonathan Gemus, "Are Older Workers Responding to the Bear Market?" Boston College's Center for Retirement Research, September 2002, available at *http://www.bc.edu/centers/crr/jtf_5.shtml.*

8. Des Dearlove, "Work Begins at Retirement," *The Times of London,* July 25, 2002. "A study last year, conducted by the HR consulting firm William M.

Mercer, Phased Retirement and the Changing Face of Retirement, is encouraging. Research among 232 US employers found that almost 60 percent had a policy for rehiring the retired."

9. Kimberly Prenda and Sidney Stahl, "The truth about older workers," *Business & Health*, May 1, 2001.
10. Art Kleiner, "Elliott Jaques Levels With You," *Strategy & Business*, issue 22, first quarter 2001; Elliott Jaques and Kathryn Cason, *Human Capability: A Study of Individual Potential and its Application*, (Gloucester, Mass: Cason Hall Publishers, 1994).
11. Richard Burkhauser, Kenneth Couch, and John Phillips, "Who Takes Early Social Security Benefits: The Economic and Health Characteristics of Early Beneficiaries," *The Gerontologist*, 1996, vol. 36, issue 6, as quoted in *Research Highlights in the Demography and Economics of Aging*, issue 3, January 1999, available at *http://agingmeta.psc.isr.umich.edu/resHigh3.pdf.*
12. "Facts and Figures," National Hospice and Palliative Care Organization, Alexandria, Virginia, August 2002, available at *http://www.nhpco.org.*
13. Bernard Starr, "Not Only is Our Society Aging, Our Prisoners are Aging as Well—and it's Costing a Fortune," *San Diego Union Tribune*, September 15, 1999.
14. Stefan Theil "Marketing to The Elder Set," *Newsweek*, September 16, 2002.

Chapter 3: The Great Flood of People

15. Kenichi Ohmae, "Profits and Perils in China, Inc.," *Strategy & Business*, issue 26, first quarter 2002.
16. John Gittings, "Growing Sex Imbalance Shocks China," *The Guardian*, May 13, 2002.
17. For further information consult the U.S. Census Bureau's Selected Historical Decennial Census Population and Housing Counts website: *http://www.census.gov/population/www/censusdata/hiscendata.html* We refer to two reports on the Web site: "United States: Urban and Rural Population: 1790 to 1990," and "1990 Population and Housing Unit Counts: United States (CPH-2)."
18. "China's Contradictions—and Possible Collapse," a Global Business Network interview with Orville Schell, September 2001, available at *http://www.gbn.org/public/gbnstory/articles/pub_chinascontradictions.htm.*
19. Kenichi Ohmae, op. cit.
20. Lexington Area Muslim Network, available at *http://leb.net/pipermail/lexington-net/2000-January/001782.html.*
21. U.S. Department of State, International Religious Freedom Report, 2002.
22. "Muslim Britain—a map of Muslim Britain," *The Guardian*, June 17, 2002.
23. U.S. Department of State, International Religious Freedom Report, 2002.

24. For more statistics refer to *http://muslim-canada.org/muslimstats.html*

25. Ambrose Evans-Pritchard, "Antwerp Race Riots Militant Charged," *The Daily Telegraph,* November 30, 2002. available at *http://www.eurozine.com/article/2000-11-15-drakulic-en.html*

26. Slavenka Drakulic, "Who Is Afraid of Europe?" opening speech for the fourteenth European Meeting of Cultural Journals, *Politics and Cultures in Europe: New Visions, New Divisions,* Vienna, November 9, 2000, available at *http://www.eurozine.com/article/2000-11-15-drakulic-en.html*

Chapter 4: The Return of the Long Boom

27. Robert Gordon, "Hi-tech Innovation and Productivity Growth: Does Supply Create its Own Demand?" December 19, 2002, available at *http://www.econ.northwestern.edu/faculty-frame.html.*

28. Anna Bernasek, "The Productivity Miracle Is For Real," *Fortune,* March 18, 2002.

29. Robert Gordon, "Two Centuries of Economic Growth: Europe Chasing the American Frontier," prepared for Economic History Workshop, Northwestern University, October 17, 2002, available at *http://www.econ.northwestern.edu/faculty-frame.html.*

30. Sources include Anna Bernasek, "The Productivity Miracle Is For Real," *Fortune,* March 18, 2002; Jerry Useem, "And Then, Just When You Thought the 'New Economy' Was Dead . . . ," *Business 2.0,* August 2001; Robert Gordon, "Hi-Tech Innovation and Productivity Growth: Does Supply Create its Own Demand?" December 19, 2002, available at *http://faculty-web.at.nwu.edu/economics/gordon/NBERPaper.pdf.*

31. Oxford University economic historian Paul David as quoted by Jerry Useem in "And Then, Just When You Thought the 'New Economy' Was Dead . . . ," *Business 2.0,* August 2001.

32. Robert Gordon, "Two Centuries of Economic Growth: Europe Chasing the American Frontier," prepared for Economic History Workshop, Northwestern University, October 17, 2002, available at *http://faculty-web.at.nwu.edu/economics/gordon/355.pdf.*

33. Peter Drucker, *Post-Capitalist Society* (New York: HarperCollins, 1993).

34. Robert Gordon, "Two Centuries of Economic Growth: Europe Chasing the American Frontier," prepared for Economic History Workshop, Northwestern University, October 17, 2002, available at *http://faculty-web.at.nwu.edu/economics/gordon/355.pdf.*

35. Joseph Stiglitz, *Globalization and its Discontents* (New York: W.W. Norton & Company, 2002).

36. Hernando de Soto, *The Mystery of Capital* (New York: Basic Books, 2000).

37. Art Kleiner, "The Next Wave of Format," published by Global Business Network, June 2001, available at *http://www.gbn.org/public/gbnstory/articles/ex_format.htm.*

38. Tim O'Reilly, "Piracy is Progressive Taxation, and Other Thoughts on the

Evolution on Online Distribution," December 12, 2002, available at *www.oreillynet.com/pub/a/p2p/2002/12/11/piracy.html.*

39. Ibid.

Chapter 5: The Thoroughly New World Order

40. Art Kleiner, "The Dilemma Doctors," *Strategy and Business,* issue 23, second quarter 2001.

41. John T. Correll, "The Evolution of the Bush Doctrine," *Air Force Magazine Online,* vol. 86, no. 02, February 2003, available at *http://www.afa.org/magazine/Feb2003/02evolution03.asp.*

42. Will Hutton, "Does Old Europe Hate New America, or Just its President?" *New York Observer,* February 24, 2003.

43. "Newsgram," *U.S. News & World Report,* January 14, 1980.

44. John Harris & Thomas Lippman, "Clinton Faces Challenges on China Policy; Pre-Summit Speech to Stress Cooperation," *The Washington Post,* October 24, 1997.

45. Lawrence Kaplan, "Guess Who Hates America? Conservatives," *The New Republic,* June 26, 2000.

46. See "A 'World-Class' Reflective Practice Field," Peter Senge et al, *The Dance of Change* (New York: Doubleday, 1999).

47. Eric Schmitt, "U.S. Combat Force of 1,700 Is Headed to the Philippines," *The New York Times,* February 21, 2003.

48. For further information about the European Union consult "The History of the European Union: A Chronology from 1946 to 2002," available at *http://europa.eu.int/abc/history/index_en.htm.*

49. Joseph Nye, *The Paradox of American Power: Why the World's Only Superpower Can't Do It Alone* (New York: Oxford University Press, 2002). Paraphrased in "What Defines National Power," a summary of a presentaion by Joseph Nye to the World Economic Forum Annual Meeting, January 30, 2001, available at *http://www.weforum.org/site/knowledgenavigator.nsf/Content/What%20Defines%20National%20Power%3F.*

Chapter 6: A Catalog of Disorder

50. Philip Jenkins, "The Next Christianity," *The Atlantic Monthly,* vol. 290, no. 3, October 2002.

51. See the Hartford Institute for Religious Research, available at *http://hirr.hartsem.edu.*

52. Michael Hout, Andrew Greely, and Melissa Wilde, "The Demographic Imperative in Religious Change in the United States," *American Journal of Sociology,* 107 (2): 468–500, 2001.

53. American Religious Identification Survey, published by the City University of New York, available at *http://www.gc.cuny.edu/studies/aris_index.htm.*

54. Philip Jenkins as quoted in "Christianity's New Center," *Atlantic Un-*

bound, available at *http://theatlantic.com/unbound/interviews/int2002-09-12.htm.*

55. Ibid.

56. Nicholas Eberstadt, "The Future of AIDS," *Foreign Affairs*, vol. 81, no. 6, November/December 2002, p.22.

57. Statistics from HardTruth about AIDS page, available at *http://hardtruth.qti.net/ThailandAIDSAwarenessPage.htm*

Chapter 7: Breakthroughs in Breaking Through: Science and Technology

58. Ron Cowen, "A Dark Force in the Universe," *Science News*, vol. 159, no. 14, April 7, 2001.

59. Arthur Koestler, *The Sleepwalkers: A History of Man's Changing Vision of the Universe* (London: Arkana, 1998).

60. Siddartha Mukherjee, "The Case for Funding Curiosity," *The New Republic*, January 21, 2002.

61. Valerie Jamison, "Carbon Nanotubes Roll On," PhysicsWeb, June 2000, available at *http://physicsweb.org/article/world/13/6/7/1.*

62. Eamonn Kelly & Pete Leyden, *What's Next? Exploring New Terrain for Business* (New York: Perseus, 2002).

63. Michael Nielson, "Rules for a Complex Quantum World," *Scientific American*, vol. 287, no. 5, November 2002.

64. For more information take a look at Stanford University's distributed computing project on protein folding Web site: *http://folding.stanford.edu.*

Chapter 8: A Cleaner, Deadlier World

65. Bjorn Lomborg, *The Skeptical Environmentalist* (New York: Cambridge University Press, 2001).

66. Ibid.

67. Katherine Mieszkowski, "Steal this Car!" Salon.com, September 4, 2002, available at *http://archive.salon.com/tech/feature/2002/09/04/woe_to_ev1/print.html.*

68. Mark Fischetti, "Why Not a 40-MPG SUV?" *Technology Review*, vol. 105, no. 9, November 2002.

69. David Talbot, "The Next Nuclear Plant," *Technology Review*, vol. 105, no. 1, January/February 2002.

70. Andrea Kalin & Jacqueline Shearer, directors, *Influenza 1918: The American Experience*, PBS Home Video, 1998.

71. Randy Shilts, *And the Band Played On: Politics, People, and the Aids Epidemic* (New York: St. Martin's Press, 1987).

72. John Lewis, *Rain of Iron and Ice* (New York: Perseus, 1996).

Chapter 9: Inevitable Strategies

73. Vernor Vinge, "Technological Singularity," presented at the VISION-21 Symposium sponsored by NASA Lewis Research Center and the Ohio Aerospace Institute, March 30–31, 1993. A version of this presentation is available at *http://singularity.manilasites.com/stories/storyReader$35.*

ACKNOWLEDGMENTS

Nearly everything I have ever done involves collaboration and this book is no different. Many people contributed to the ideas and information and challenged my thinking. I hope I have been able to capture their contributions here.

Most important has been my collaborator, Art Kleiner. Though the book originated with me, Art's contributions, as always, were profound. The quality of the writing and the structure of the book are far better thanks to Art's hard work and editorial skills. But equally important he has added to and enriched many of the ideas in this book. And for this contribution I am very grateful.

My colleagues in Global Business Network made many contributions, particularly our creative demographer Chris Ertel, my researchers Joe McCrossen, Chris Coldeway, and Erik Smith, and my cofounders Stewart Brand, Jay Ogilvy, and Napier Collyns. Others at GBN (now part of the Moniter Group) also contributed, including Eamonn Kelly, Katherine Fulton, Doug Randall, and Jim Cutler. Nancy Murphy, as always, has helped make this a more readable book. And finally my assistant Laura Panica has with good cheer helped bring order to the chaos and made us all feel good about it.

My colleagues at Alta Partners have also been instrumental in my learning, especially about information technology, life sciences, and how companies are born and built. Especially important is Garrett Gruener, who has been a close friend and never fails to ask the hard questions. My partners include Jean Deleage, Dan Janney, Guy Nohra, Alix Marduel, Farah Champsi, Khaled Nasr, Robert Simon, and Ed Penhoet.

I have had the oppurtunity to work on many projects in the last few years that have been supported by a number of organizations and where I have found remarkable collaborators in the arena of national security and geopolitics. These include Andy Marshall in the office of the U.S. Secretary of Defense, Dick O'Neil of the Highlands Group, Carol Dumaine and Jim Harris at the CIA, Mike Goldblatt and Stu Wolf at DARPA, Chuck Boyd at the Hart-Rudman Commission, Shaun Jones, M.D., at the NSA, and Peter Ho at the Singapore Ministry of Defense.

My colleagues at the Long Foundation have helped me see the surprising connections between today and the deep future. Stewart Brand, Alexander Rose, Danny Hillis, Brian Eno, Esther Dyson, Paul Saffo, Roger Kennedy, Mike Keller, Mitch Kapor, and Kevin Kelly have all been part of this rich conversation. Walter Parkes, Laurie McDonald, and Steven Spielberg of Dreamworks have given me the opportunity to think about how all these ideas become manifest in the futures their films portray.

My associates at the World Business Council on Sustainable Development, the Pew Center for Climate Change, the California Energy Commission, the California Air Resources Board, and the California Environmental Dialog have given me the opportunity to participate in farsighted research on the long-term environmental future and the technologies that can help save the Earth's ecosystems.

I also participate in a rich intellectual community that informs and always challenges my thinking. I am especially grateful to Orville Schell, Chris Anderson, Paul Hawken, David Harris, Peter Calthorpe, Joel Hyatt, Louis Rosetto, Jane Metcalfe, and Nat Goldhaber.

And of course nothing would have happened without my literary agent, John Brockman, and his partner, Katinka Matson. As usual, John set me on the right course. I had planned to write a very different book but John said that no one would read it and instead, I should write about what I was really interested in. *Inevitable Surprises* is the result. At Gotham Books, my publisher Bill Shinker, my editor Brendan Cahill, and my copy editor Craig Schneider have shown remarkable patience and great skill in bringing this book to a timely fruition.

WINNER OF THE PULITZER PRIZE FOR BIOGRAPHY

KHRUSHCHEV

The Man and His Era

William Taubman

A magisterial, definitive and compelling assessment of one of the giants of twentieth-century history: former Soviet leader Nikita Khrushchev.

'Outstanding, superbly gripping and surely definitive . . . Fascinating'
– Simon Sebag Montefiore, DAILY TELEGRAPH

'[A] monumental biography, one that is likely to be definitive for years to come' – Leon Aron, NEW YORK TIMES

'Masterly . . . Taubman has painted a remarkable portrait of a politician' – Richard Overy, SUNDAY TELEGRAPH

'A monumental book . . . A masterpiece, magnificently researched and well written, bringing out the true dimensions of his subject' – Simon Heffer, SPECTATOR

PRICE £14.99
ISBN 0-7432-3166-X

THE STASI FILES

East Germany's Secret Operations Against Britain

Anthony Glees

The definitive, revelatory account of the GDR's Cold War espionage activities against the UK: their goals, methods, sources – and recruits.

'Brilliantly illustrates the weird, horrifying, viciously cruel place that was Cold War East Germany'
– Andrew Roberts, EVENING STANDARD

Before the collapse of Communist East Germany the country ran one of the most extensive intelligence networks in the world. The Stasi was a highly professional and ruthless organisation which was dedicated to principles of conspiratorial aggressiveness and the protection of the Communist cause.

Revelatory and controversial, *The Stasi Files* is the most important book on espionage to appear since *The Mitrokhin Archive.*

PRICE £9.99
ISBN 0-7432-3105-8

ONE NO, MANY YESES

A Journey to the Heart of the Global Resistance Movement

Paul Kingsnorth

A manifesto, an investigation, a travel book: an introduction to the new politics of resistance which shows there's much more to the anti-globalisation movement than trashing Starbucks.

'Gripping, engaging and inspiring – it will become a classic'
GEORGE MONBIOT

It could turn out to be the biggest political movement of the twenty-first century: a global coalition of millions, united in resisting an out-of-control global economy, and already building alternatives to it. Not socialism, not capitalism, not any 'ism' at all, it is united in what it opposes, and deliberately diverse in what it wants instead – a politics of 'one no, many yeses'. This movement may change the world. This book tells its story.

PRICE £7.99
ISBN 0-7432-2027-7

BLAIR'S WARS

John Kampfner

A riveting and thought-provoking insight into the processes by which Tony Blair has taken us to war more often than any other recent Prime Minister.

'Brilliant . . . highly perceptive' – Anthony Howard, SUNDAY TIMES Books of the Year

'A brilliant book by one of Britain's most distinguished political writers' – MAIL ON SUNDAY

'Kampfner has some excellent sources and there is at least one revealing moment every few pages . . . [A] must-read book about New Labour' – INDEPENDENT ON SUNDAY

'The most perceptive book about the Blair government to appear since Andrew Rawnsley's *Servants of the People*' – SUNDAY TIMES

PRICE £7.99

ISBN 0-7432-4830-9

THE BASE

Al-Qaeda and the Changing Face of Global Terror

Jane Corbin

The definitive investigation into the background, personnel and methods of al-Qaeda.

Jane Corbin has been studying Bin Laden's organisation for four years and has followed in his footsteps through the Middle East, Africa, Europe and America. She has conducted hundreds of interviews with key eyewitnesses, investigators and intelligence officers around the world. Tracing al-Qaeda's roots back to the jihad against the Soviets in Afghanistan, Jane Corbin picks up the complicated trail that led to the collapse of the Twin Towers and beyond.

The Base is essential reading for anyone interested in the history and likely future operations of arguably the biggest threat to democracy since the Cold War.

PRICE £8.99

ISBN 0-7434-4942-8

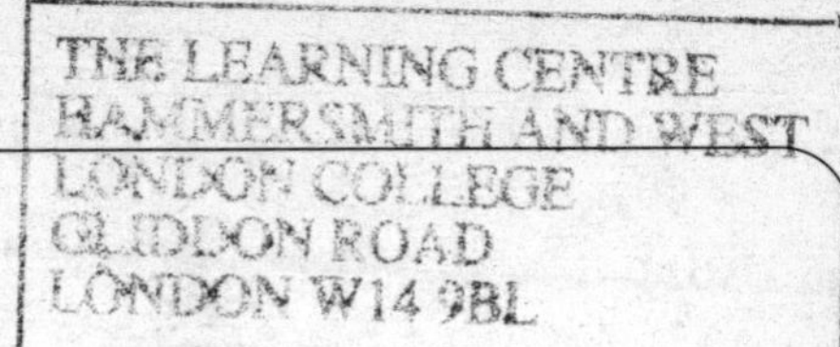

SIMON &
SCHUSTER